ELECTRONICS
FOR
ELECTRICIANS

ELECTRONICS
FOR
ELECTRICIANS

TREVOR LINSLEY

Senior Lecturer, Blackpool and The Fylde College

Edward Arnold

A division of Hodder & Stoughton

LONDON   MELBOURNE   AUCKLAND

© 1990 TREVOR LINSLEY

First published in Great Britain 1990
Reprinted 1992 (with corrections), 1993

British Library Cataloguing in Publication Data
Linsley, Trevor
   Electronics for electricians.
   1. Electronics
   I. Title
   621.381

   ISBN 0-340-52511-8

Typeset in Times Roman by Keyset Composition, Colchester, Essex.
Printed and bound in Great Britain for Edward Arnold,
a division of Hodder and Stoughton Limited, Mill Road, Dunton
Green, Sevenoaks, Kent TN13 2YA by Thomson Litho Ltd,
East Kilbride, Scotland.

To Joyce, Samantha and Victoria

PREFACE

This book covers the electronic content of all the City and Guilds Electrical Installation courses. In particular it provides a text book for the new Basic Installation Electronics Technology content of the 236 Electrical Installation Competences, the City and Guilds Supplementary Studies in Electronics and the 232 Electrical and Electronic Craft Studies course.

In addition to meeting the specific requirements of these courses, it is also hoped that the book will provide a basic 'user friendly' guide to electronics for enthusiasts and students on BTEC and other City and Guilds courses.

Electronics has today found its way into most industrial applications and consequently many craftsmen now find that a basic knowledge of electronics is essential to carry out their work efficiently. One of my aims in writing this book is to provide readers with a basic working knowledge of electronics which they can quickly apply to their own practical situation. The Appendix brings together some of the basic reference information which those new to electronics will find useful.

The treatment of electronics in this book is a non-mathematical one. However, for those who require a deeper understanding of electronic circuits, I have included chapter three, Circuit Theory, which should perhaps be 'dipped into' on a 'need to know' basis.

Electronics has created many new opportunities for electricians, technicians and installers and for this reason I have included the chapter on security systems, communication systems and computer networks. Those involved in process control may find the chapter on transducers and electronic systems particularly interesting.

I would like to acknowledge the assistance given by the following manufacturers and organisations in the preparation of this book:

Crabtree Electrical Industries Ltd.
Farnell Components Ltd.
Meggar Instruments Ltd. (AVO)
M.K. Electric Ltd.
Multicore Solders Ltd.
R.S. Components Ltd.

I would like to acknowledge my gratitude to the Open University for my own electronics education, the proposal reviewers and my colleagues at Blackpool and the Fylde College for their suggestions and assistance during the preparation of the manuscript.

Finally, I would like to thank Joyce, Samantha and Victoria for their support and encouragement.

Trevor Linsley,
Poulton-le-Fylde   1990.

CONTENTS

CHAPTER 1

Electronic component recognition

There are numerous electronic components, diodes, transistors, thyristors and integrated circuits each with its own limitations, characteristics, and designed application. When repairing electronic circuits it is important to replace damaged components with an identical or equivalent component. Manufacturers issue comprehensive catalogues with details of working voltage, current, power dissipation etc., and the reference numbers of equivalent components, and some of this information is included in the Appendix. These catalogues of information, together with a high impedance multi-meter as described in Chapter 5, should form a part of the extended tool kit for an installation electrician proposing to repair electronic circuits.

Electronic circuit symbols

The British Standard BS 3939 recommends that particular graphical symbols be used to represent a range of electronic components on circuit diagrams. The same British Standard recommends a range of symbols suitable for electrical installation circuits with which electricians will already be familiar. Figure 1.1 shows a selection of electronic symbols.

Resistors

All materials have some resistance to the flow of an electric current but, in general, the term *resistor* describes a conductor specially chosen for its resistive properties.

Resistors are the most commonly used electronic component and they are made in a variety of ways to suit the particular type of application. They are usually manufactured as either carbon composition or carbon film. In both cases the base resistive material is carbon and the general appearance is of a small cylinder with leads protruding from each end, as shown in Figure 1.2(a).

If subjected to overload, carbon resistors usually decrease in resistance since carbon has a negative temperature coefficient. This causes more current to flow through the resistor, the temperature rises and failure occurs, usually by fracturing. Carbon resistors have a power rating of between 0.1 W and 2 W which should not be exceeded.

When larger power rated resistors are required a wire wound resistor should be chosen. This consists of a resistance wire of known value wound on a small ceramic cylinder which is encapsulated in a vitreous enamel coating as shown in Figure 1.2(b). Wire wound resistors are designed to run hot and have a power rating up to 20 W. Care should be taken when mounting wire wound resistors to prevent the high operating temperature affecting any surrounding components.

A variable resistor is one which can be varied continuously from a very low value to the full rated resistance. This characteristic is required in tuning circuits to adjust the signal or voltage level for brightness, volume or tone. The most common type used in electronic work has a circular carbon track contacted by a metal wiper arm. The wiper arm can be adjusted by means of an adjusting shaft (rotary type) or by placing a screwdriver in a slot (preset type) as shown in Figure 1.3. Variable resistors are also known as potentiometers because they can be used to adjust the potential difference (voltage) in a circuit. The variation in resistance can be either to a logarithmic or linear scale.

The value of the resistor and the tolerance may be marked on the body of the component either by direct numerical indication or by using a standard colour code. The method used will depend upon

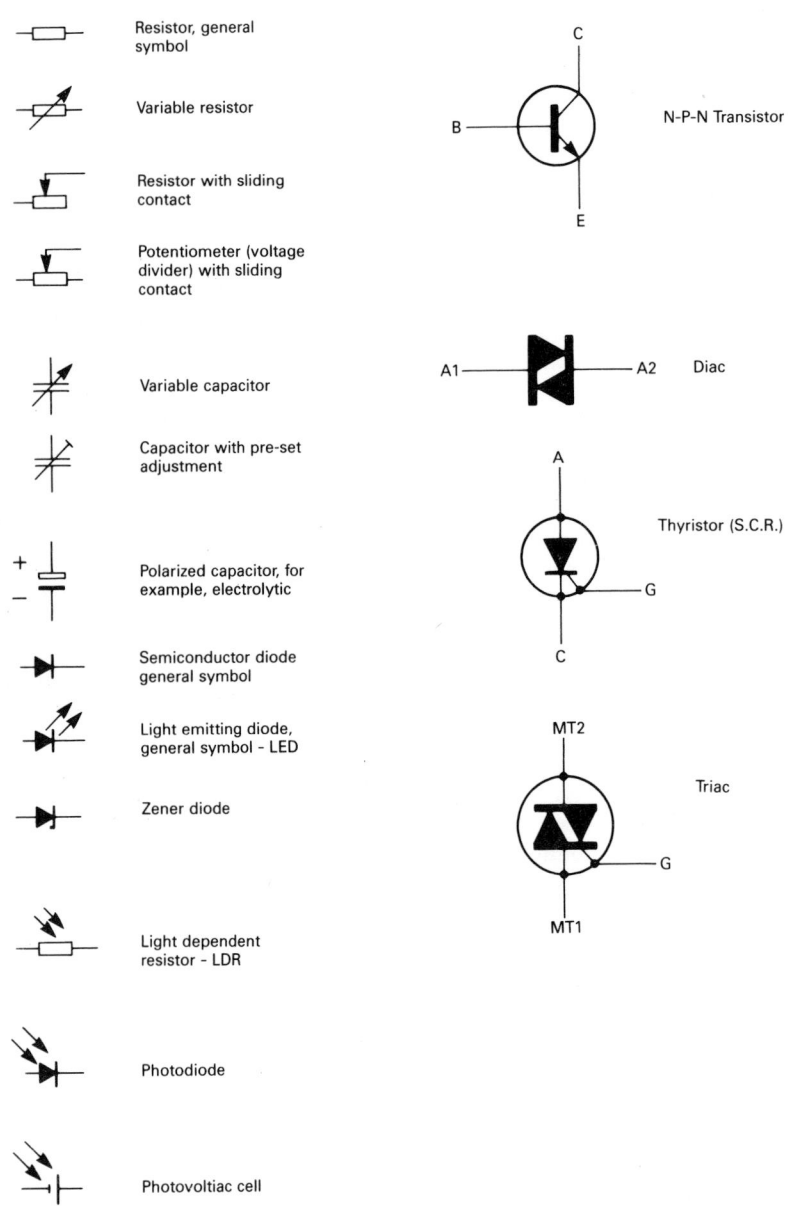

Resistor, general symbol

Variable resistor

Resistor with sliding contact

Potentiometer (voltage divider) with sliding contact

Variable capacitor

Capacitor with pre-set adjustment

Polarized capacitor, for example, electrolytic

Semiconductor diode general symbol

Light emitting diode, general symbol – LED

Zener diode

Light dependent resistor – LDR

Photodiode

Photovoltiac cell

N-P-N Transistor

Diac

Thyristor (S.C.R.)

Triac

Figure 1.1 Some BS 3939 graphical symbols used in electronics

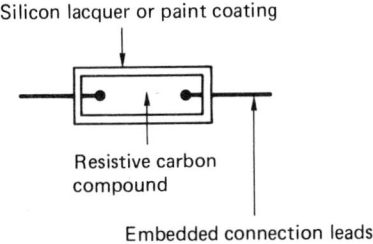

Silicon lacquer or paint coating

Resistive carbon compound

Embedded connection leads

(a) Carbon composition resistor

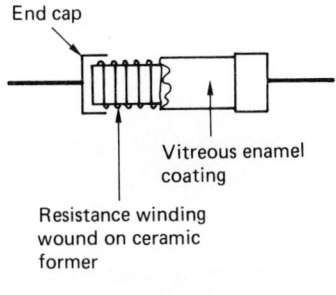

End cap

Vitreous enamel coating

Resistance winding wound on ceramic former

(b) Wire wound resistor

Figure 1.2 Construction of resistors

the type, physical size and manufacturer's preference, but in general the larger components have values marked directly on the body and the smaller components use the standard resistor colour code.

Abbreviations used in electronics

Where the numerical value of a component includes a decimal point, it is standard practice to include the prefix for the multiplication factor in place of the decimal point, to avoid accidental marks being mistaken for decimal points. Multiplication factors and prefixes are dealt with in Chapter 3.

The abbreviation R means $\times$ 1
k means $\times$ 1 000
M means $\times$ 1 000 000

Therefore, a 4.7 KΩ resistor would be abbreviated to 4 k7. A 5.6 Ω resistor to 5R6 and a 6.8 MΩ resistor to 6M8.

Tolerances may be indicated by adding a letter at the end of the printed code.

The abbreviation F means $\pm$ 1%
G means $\pm$ 2%
J means $\pm$ 5%
K means $\pm$ 10%
M means $\pm$ 20%

Therefore a 4.7 kΩ resistor with a tolerance of 2% would be abbreviated to 4k7G. A 5.6 Ω resistor with a tolerance of 5% would be abbreviated to 5R6J. A 6.8 MΩ resistor with a 10% tolerance would be abbreviated to 6M8K.

This is the British Standard BS 1852 code which is recommended for indicating the values of resistors on circuit diagrams and components when their physical size permits.

The standard colour code

Small resistors are marked with a series of coloured bands as shown in Table 1.1. These are read according to the standard colour code to determine the resistance. The bands are located on the component towards one end. If the resistor is turned so that this end is towards the left, the

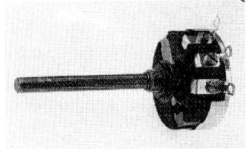

Rotary type

Preset type

Figure 1.3 Types of variable resistor

Table 1.1   The resistor colour code

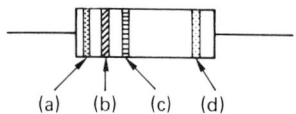

Colour	Band (a) first number	Band (b) second number	Band (c) number of noughts	Band (d) tolerance band
Black	0	0	None	–
Brown	1	1	1	1%
Red	2	2	2	2%
Orange	3	3	3	–
Yellow	4	4	4	–
Green	5	5	5	–
Blue	6	6	6	–
Violet	7	7	7	–
Grey	8	8	–	–
White	9	9	–	–
Gold	–	–	÷10	5%
Silver	–	–	÷100	10%
None	–	–	–	20%

bands are then read from left to right. Band (a) gives the first number of the component value, band (b) the second number, band (c) the number of zeros to be added after the first two numbers and band (d) indicates the resistor tolerance, which is commonly gold or silver, indicating a tolerance of 5% or 10% respectively. If the bands are not clearly oriented towards one end, first identify the tolerance band and turn the resistor so that this is towards the right before commencing to read the colour code as described.

The tolerance band indicates the maximum tolerance variation in the declared value of resistance. Thus a 100 Ω resistor with a 5% tolerance, will have a value somewhere between 95 Ω and 105 Ω since 5% of 100 Ω is 5 Ω.

Example 1

A resistor is colour coded yellow, violet, red, gold. Determine the value of the resistor.

band (a) – yellow has a value of 4,
band (b) – violet has a value of 7,
band (c) – red has a vlue of 2,
band (d) – gold indicates a tolerance of 5%.

The value is therefore 4 700 ± 5%
This could be written as 4.7 kΩ ± 5% or 4k7J.

Example 2

A resistor is colour coded green, blue, brown, silver. Determine the value of the resistor.

band (a) – green has a value of 5,
band (b) – blue has a value of 6,
band (c) – brown has a value of 1,
band (d) – silver indicates a tolerance of 10%.

The value is therefore 560 ± 10% and could be written as 560 Ω ± 10% or 560RK.

Example 3

A resistor is colour coded blue, grey, green, gold. Determine the value of the resistor.

band (a) – blue has a value of 6,
band (b) – grey has a value of 8,
band (c) – green has a value of 5,
band (d) – gold indicates a tolerance of 5%.

The value is therefore 6 800 000 ± 5% and could be written as 6.8 MΩ ± 5% or 6M8J.

Example 4

A resistor is colour coded orange, white, silver, silver. Determine the value of the resistor.

band (a) – orange has a value of 3,
band (b) – white has a value of 9,
band (c) – silver indicates divide by 100 in this band,
band (d) – silver indicates a tolerance of 10%.

The value is therefore 0.39 ± 10% and could be written as 0.39 Ω ± 10% or R39K.

Preferred values

It is difficult to manufacture small electronic resistors to exact values by mass production methods. This is not a disadvantage as in most electronic circuits the value of the resistors is not critical. Manufacturers produce a limited range of *preferred* resistance values rather than an over-whelming number of individual resistance values. Therefore, in electronics, we use the preferred value closest to the actual value required.

A resistor with a preferred value of 100 Ω and a

Table 1.2 Preferred values

E6 series 20% tolerance	E12 series 10% tolerance	E24 series 5% tolerance
10	10	10
		11
	12	12
		13
15	15	15
		16
	18	18
		20
22	22	22
		24
	27	27
		30
33	33	33
		36
	39	39
		43
47	47	47
		51
	56	56
		62
68	68	68
		75
	82	82
		91

10% tolerance could have any value between 90 Ω and 110 Ω. The next larger preferred value which would give the maximum possible range of resistance values without too much overlap would be 120 Ω. This could have any value between 108 Ω and 132 Ω. Therefore, these two preferred value resistors cover all possible resistance values between 90 Ω and 132 Ω. The next preferred value would be 150 Ω, then 180 Ω, 220 Ω and so on.

There is a series of preferred values for each tolerance level as shown in Table 1.2 so that every possible numerical value is covered. Table 1.2 indicates values between 10 and 100 but larger values can be obtained by multiplying these preferred values by a multiplication factor. Resistance values of 47 Ω, 470 Ω, 4.7 kΩ, 470 kΩ, 4.7 MΩ etc., are available in this way.

Testing resistors

The resistor being tested should have a value close to the preferred value and within the tolerance stated by the manufacturer. To measure the resistance of a resistor which is not connected into a circuit, the leads of a suitable ohm meter should be connected to each resistor connection lead and a reading obtained. The ohm meter and its use are discussed in Chapter 5.

If the resistor to be tested is connected into an electronic circuit it is *always necessary*, first to disconnect one lead from the circuit before the test leads are connected, otherwise the components in the circuit will provide parallel paths, and an incorrect reading will result.

Capacitors

The fundamental principles of capacitors and the time constant of capacitor resistor circuits are discussed in Chapter 3 under the sub-heading *Electrostatics*. In this chapter we shall consider the practical aspects associated with capacitors in electronic circuits.

A capacitor stores a small amount of electric charge; it can be thought of as a small rechargeable battery which can be quickly recharged. In electronics we are not only concerned with the amount of charge stored by the capacitor but in the way the value of the capacitor determines the performance of timers and oscillators by varying the time constant of C-R circuits.

Capacitors in action

If a test circuit is assembled as shown in Figure 1.4 and the changeover switch connected to d.c. the signal lamp will only illuminate for a very short pulse as the capacitor charges. The charged capacitor then blocks any further d.c. current flow as shown by the graphs of Figure 3.4. If the changeover switch is then connected to a.c. the lamp will

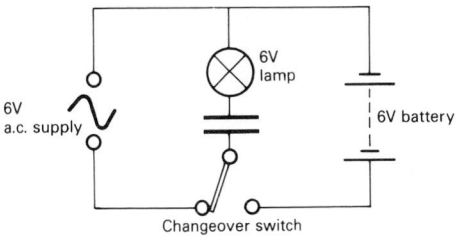

Figure 1.4 Test circuit showing capacitors in action

illuminate at full brilliance because the capacitor will charge and discharge continuously at the supply frequency. Current is *apparently* flowing through the capacitor because electrons are moving to and fro in the wires joining the capacitor plates to the a.c. supply.

Coupling and decoupling capacitors

Capacitors can be used to separate a.c. and d.c. in an electronic circuit. If the output from circuit A shown in Figure 1.5(a) contains both a.c. and d.c. but only an a.c. input is required for circuit B then a *coupling* capacitor is connected between them. This blocks the d.c. while offering a low reactance to the a.c. component. Alternatively, if it is required that only d.c. be connected to circuit B shown in Figure 1.5(b), a *decoupling* capacitor can be connected in parallel with circuit B. This will provide a low reactance path for the a.c. component of the supply and only d.c. will be presented to the input of B. This technique is used to *filter out* unwanted a.c. in, for example, d.c. power supplies.

Types of capacitor

There are two broad categories of capacitor, the non-polarised and polarised type. The non-polarised type can be connected either way round but polarised capacitors *must* be connected to the polarity indicated otherwise a short circuit and consequent destruction of the capacitor will result. There are many different types of capacitor, each

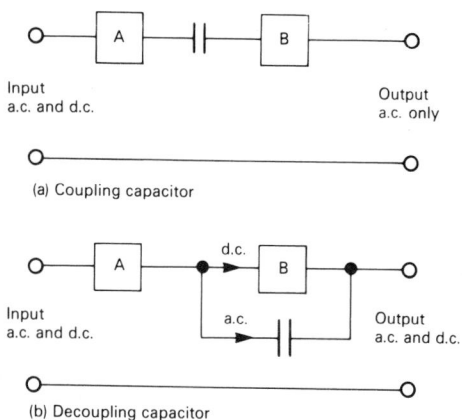

(a) Coupling capacitor

(b) Decoupling capacitor

Figure 1.5 Coupling and decoupling capacitors

one being distinguished by the type of dielectric used in its construction. Figure 1.6 shows some of the capacitors used in electronics.

Polyester capacitors

Polyester capacitors are an example of the plastic film capacitor. Polypropylene, polycarbonate and polystyrene capacitors are other types of plastic film capacitors. The capacitor value may be marked on the plastic film or the capacitor colour code given in Table 1.3 may be used. This dielectric material gives a compact capacitor with good electrical and temperature characteristics. They are used in many electronic circuits but are not suitable for high-frequency use.

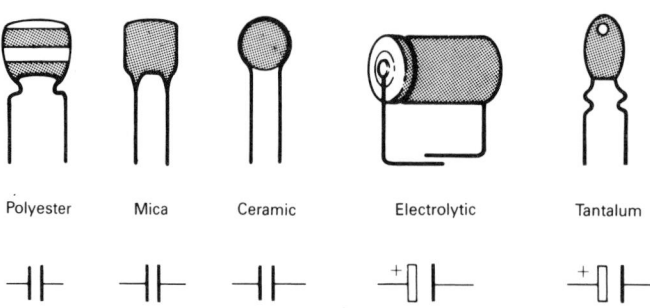

Figure 1.6 Capacitors and their symbol used in electronic circuits

Table 1.3   Colour code for plastic film capacitors

Plastic film
series C280
capacitors

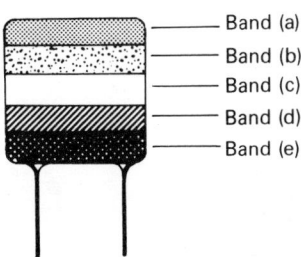

Band (a)
Band (b)
Band (c)
Band (d)
Band (e)

| Capacitor colour code for plastic film capacitors. Values in pF (picofarad − ×10^{-12}F) | | | | |
Colour	Band (a) first number	Band (b) second number	Band (c) number of noughts to be added	Band (d) tolerance	Band (e) maximum voltage
black	–	0	None	20%	–
brown	1	1	1	–	100 V
red	2	2	2	–	250 V
orange	3	3	3	–	–
yellow	4	4	4	–	400 V
green	5	5	5	5%	–
blue	6	6	6	–	–
violet	7	7	–	–	–
grey	8	8	–	–	–
white	9	9	–	10%	–

Mica capacitors

Mica capacitors have excellent stability and are accurate to ± 1% of the marked value. Since costs usually increase with increased accuracy, they tend to be more expensive than plastic film capacitors. They are used where high stability is required, for example, in tuned circuits and filters.

Ceramic capacitors

Ceramic capacitors are mainly used in high-frequency circuits subjected to wide temperature variations. They have a high stability and low loss.

Electrolytic capacitors

Electrolytic capacitors are used where a large value of capacitance coupled with a small physical size is required. They are constructed on the 'Swiss roll' principle as are the paper dielectric capacitors used for p.f. correction in electrical installation circuits. The electrolytic capacitors' high capaci-tance for very small volume is derived from the extreme thinness of the dielectric coupled with a high dielectric strength. Electrolytic capacitors have a size gain of approximately one hundred times over the equivalent non-electrolytic type. Their main disadvantage is that they are polarised and must be connected to the correct polarity in a circuit. Their large capacity makes them ideal as smoothing capacitors in power supplies.

Tantalum capacitors

Tantalum capacitors are a new type of electrolytic capacitor using tantalum and tantalum oxide to give a further capacitance/size advantage. They look like a 'raindrop' or 'blob' with two leads protruding from the bottom. The polarity and values may be marked on the capacitor or the colour code shown in Table 1.4 may be used. The voltage ratings available tend to be low as with all electrolytic capacitors. They are also extremely vulnerable to reverse voltages in excess of 0.3 V. This means that even when testing with an ohm

Table 1.4   Colour code for tantalum polarised capacitors

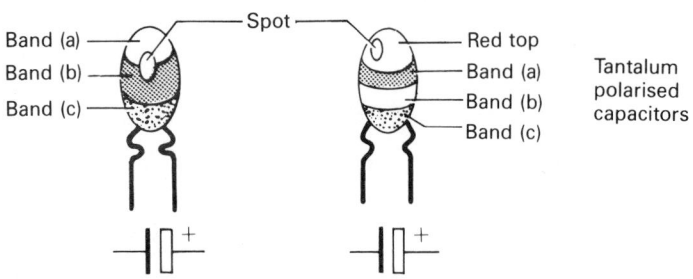

Capacitor colour code for tantalum capacitors. Values in μF (microfarad $- \times 10^{-6}$F)			
Colour	Band (a) first number	Band (b) second number	Spot number of noughts to be added	Band (c) maximum voltage
black	–	0	None	10 V
brown	1	1	1	–
red	2	2	2	–
orange	3	3	–	–
yellow	4	4	–	6.3 V
green	5	5	–	16 V
blue	6	6	–	20 V
violet	7	7	–	–
grey	8	8	$\div$100	25 V
white	9	9	$\div$1000	30 V
pink				35 V

meter, extreme care must be taken to ensure correct polarity.

Variable capacitors

Variable capacitors are constructed so that one set of metal plates move relative to another set of fixed metal plates as shown in Figure 1.7. The plates are separated by air or mica sheet which acts as the dielectric. Air dielectric variable capacitors are used to tune radio receivers to a chosen station and small variable capacitors called *trimmers* or *presets* are used to make fine, infrequent adjustments to the capacitance of a circuit.

Selecting a capacitor

When choosing a capacitor for a particular application, three factors must be considered, value, working voltage and leakage current.

The unit of capacitance is the *farad* (symbol F) to commemorate the name of the English scientist Michael Faraday. However, for practical purposes the farad is much too large and in electrical installation work and electronics we use fractions of a farad as follows:

1 microfarad = 1 μF = 1 $\times$ 10^{-6} F
1 nanofarad = 1 nF = 1 $\times$ 10^{-9} F
1 picofarad = 1 pF = 1 $\times$ 10^{-12} F

The p.f. correction capacitor used in a domestic fluorescent luminaire would typically have a value of 8 μF at a working voltage of 400 V. In an electronic filter circuit a typical capacitor value might be 100 pF at 63 V.

One microfarad is one million times greater than one picofarad. It may be useful to remember that

1000 pF = 1 nF, and 1000 nF = 1 μF.

The working voltage of a capacitor is the *maximum*

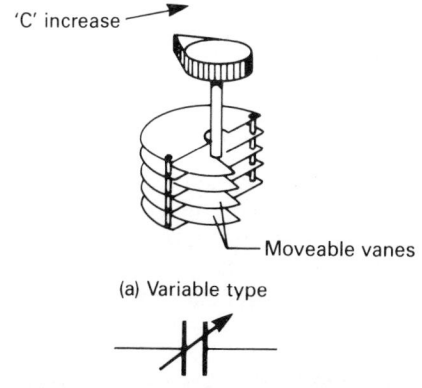

'C' increase

Moveable vanes

(a) Variable type

(b) Trimmer or preset type

Figure 1.7 Variable capacitors and their symbol

voltage that can be applied between the plates of the capacitor without breaking down the dielectric insulating material. This is a d.c. rating and, therefore, a capacitor with a 200 V rating must only be connected across a maximum of 200 V d.c. Since a.c. voltages are usually given as r.m.s. values, a 200 V a.c. supply would have a maximum value of about 283 V which would damage the 200 V capacitor. When connecting a capacitor to the 240 V mains supply we must choose a working voltage of about 400 V because 240 V r.m.s. is approximately 340 V maximum. The 'factor of safety' is small and, therefore, the working voltage of the capacitor must not be exceeded.

An ideal capacitor which is isolated will remain charged forever, but in practice no dielectric insulating material is perfect, and the charge will slowly *leak* between the plates, gradually discharging the capacitor. The loss of charge by leakage through it should be very small for a practical capacitor.

Capacitor colour code

The actual value of a capacitor can be identified by using the colour codes given in Tables 1.3 and 1.4 in the same way that the resistor colour code was applied to resistors.

Example 1

A plastic film capacitor is colour coded from top to bottom, brown, black, yellow, black, red. Deter-

mine the value of the capacitor, its tolerance and working voltage.

band (a) – brown has a value 1
band (b) – black has a value 0
band (c) – yellow indicates multiply by 10,000
band (d) – black indicates 20%
band (e) – red indicates 250 V

The capacitor has a value of 100,000 pF or 0.1 μF with a tolerance of 20% and a maximum working voltage of 250 V.

Example 2

Determine the value, tolerance and working voltage of a polyester capacitor colour coded from top to bottom, yellow, violet, yellow, white, yellow.

band (a) – yellow has a value 4
band (b) – violet has a value 7
band (c) – yellow indicates multiply by 10,000
band (d) – white indicates 10%
band (e) – yellow indicates 400 V.

The capacitor has a value of 470,000 pF or 0.47 μF with a tolerance of 10% and a maximum working voltage of 400 V.

Example 3

A plastic film capacitor has the following coloured bands from its top down to the connecting leads, blue, grey, orange, black, brown. Determine the value, tolerance and voltage of this capacitor.

band (a)   –   blue has a value 6
band (b)   –   grey has a value 8
band (c)   –   orange indicates multiply by 1000
band (d)   –   black indicates 20%
band (e)   –   brown indicates 100 V.

The capacitor has a value of 68,000 pF or 68 nF with a tolerance of 20% and a maximum working voltage of 100 V.

Capacitance value codes

Where the numerical value of the capacitor includes a decimal point, it is standard practice to use the prefix for the multiplication factor in place of the decimal point. This is the same practice which we used earlier for resistors.

The abbreviation μ means microfarad
                   n means nanofarad
                   p means picofarad

Therefore, a 1.8 pF capacitor would be abbreviated to 1p8. A 10 pF capacitor to 10p. A 150 pF capacitor to 150p or n15. A 2,200 pF capacitor to 2n2 and a 10,000 pF capacitor to 10n.

1000 pF   =   1 nF   =   0.001 μF.

Testing capacitors

The discussion earlier about *ideal* and *leaky* capacitors provides us with a basic principle to test for a faulty capacitor.

Non-polarised capacitors

Using an ohm meter as described in Chapter 5, connect the leads of the capacitor to the ohm meter and observe the reading. If the resistance is less than about 1 MΩ, it is allowing current to pass from the ohm meter and, therefore, the capacitor

is leaking and is faulty. With large-value capacitors (in the μF range) there may be a short initial burst of current as the capacitor charges up.

Polarised capacitors

It is essential to connect the *true positive* of the ohm meter to the positive lead of the capacitor. When first connected, the resistance is low but rises to a steady value as the dielectric forms between the capacitor plates.

Inductors and transformers

An inductor is a coil of wire wound on a former having a core of air or iron. When a current flows through the coil a magnetic field is established. A transformer consists of two coils wound on a common magnetic core and, therefore, in this sense, the transformer is also an inductor. Simple transformer theory is discussed in Chapter 3. A small electronic transformer and the aerial of a radio receiver comprising a coil wound on a ferrite core is shown in Figure 1.8.

Inductors such as the radio receiver aerial can be connected in parallel with a variable capacitor and *tuned* for maximum response so that a particular radio station can be listened to while excluding all others.

Most electronic circuits require a voltage between 5 and 12 volts and the transformer provides an ideal way of initially reducing the mains voltage to a value which is suitable for the particular electronic circuit.

When compared with other individual electronic components, inductors are large. The magnetic fields produced by industrial electronic equipment such as electromagnets, relays and transformers

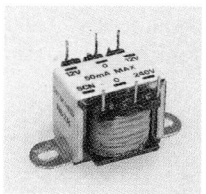

(a) Transformer

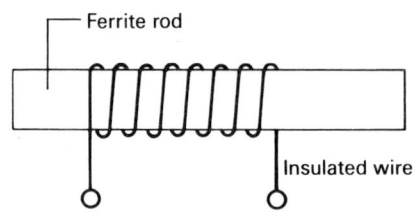

(b) Radio receiver aerial

Figure 1.8 Examples of an inductor

can cut across other electronic components and cause undesirable emf's to be induced. This causes electrical interference – called *electrical noise*, and may prevent the normal operation of the electronic circuit. This interference can be avoided by magnetically *screening* the inductive components from the remaining circuits.

Switches

In electrical installation work we identify four separate types of switching; switching for isolation, mechanical maintenance, emergency switching and functional switching. In electronics we are principally concerned with functional switching which is defined as the switching of electrically operated equipment in normal service. In any switch, metal contacts are brought together or separated by the action of the switch either to make or break the circuit. Figure 1.9 shows examples of some switches used in electronics.

Manufacturers' technical specifications rate switches according to their current rating, maximum working voltage and type of supply. The switch chosen for a particular application must be capable of interrupting the total steady current. This is the current rating quoted for most switches and relays and usually assumes a purely resistive load. If the load is inductive the switch must be derated by something between 25% and 50% if the safe working life of the switch is to be maintained.

The voltage rating indicates the maximum safe working voltage and is determined by the type of insulation material used in the construction of the switch and the contact separation. Again a purely

resistive load is assumed, and because inductive loads may cause high voltages and current surges as the magnetic flux collapses, the switch must be derated when used with inductive loads to prevent flash-over damaging the insulation and reducing the life of the switch.

The arc established across a switch contact tends to extinguish itself every half cycle as the voltage falls to zero on an a.c. supply. A d.c. arc tends to maintain itself until contact separation becomes too great. For this reason switch ratings are usually lower when used on d.c. supplies.

At voltages below about 12 V the current rating of the switch can usually be increased without seriously reducing its working life. Switch life is usually based upon a minimum of 10,000 operations at the maximum rating for a purely resistive load. This covers both electrical and mechanical wear.

Switches have a different number of contact configurations known as poles and throws as shown in Table 1.5. The poles (P) are the number of separate circuits which the switch makes or breaks at the same time. The throws (T) are the number of positions to which each pole can be switched. A one-way lighting switch allows us to change between two stable states, on and off, and is therefore called a single-pole single-throw (S.P.S.T.) switch. A two-way lighting switch moves a single pole between two contacts and is, therefore, called a single-pole, double-throw (S.P.D.T.) switch or changeover switch. A single switch toggle can be made to move two poles at the same time leading to double-pole single-throw (D.P.S.T.) and double-pole, double-throw (D.P.D.T.) switches.

(a) Toggle switch

(b) Slide switch

Figure 1.9 Switches used in electronics

Table 1.5   Switch contact configurations

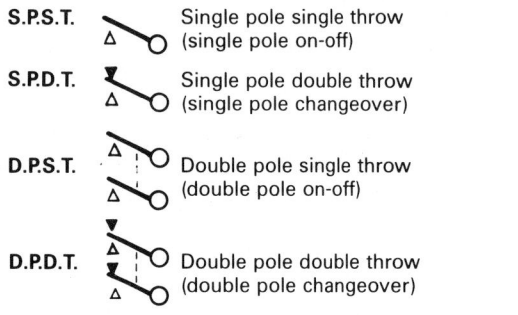

S.P.S.T.		Single pole single throw (single pole on-off)
S.P.D.T.		Single pole double throw (single pole changeover)
D.P.S.T.		Double pole single throw (double pole on-off)
D.P.D.T.		Double pole double throw (double pole changeover)

When using switches for the first time it is wise to use an ohm meter to trace out which contacts are joined when the switch is operated.

Microswitches

A microswitch is a small sensitive mechanical switch usually fitted with a lever or actuator so that only a small force is required to operate a snap action switch. The actuator may be a simple lever or incorporate a roller as shown in Figure 1.10. The actuator causes contacts in the switch to open or close in various switch configurations from S.P.S.T. to D.P.D.T.

Reed switches

A reed switch has two thin strips of steel (reeds) sealed inside a glass tube usually containing nitrogen or some other chemically inert gas to reduce sparking at the contacts, as shown in Figure 1.11. The reeds are arranged so that their ends overlap but are a short distance apart. If a magnet is brought close to the reed switch the contacts attract each other and make electrical contact, closing the switch. When the magnet is removed the reeds separate and the circuit is broken. The reed switch is a *proximity* switch since it is operated by the nearness of the magnet. It can operate very quickly, 2000 times per minute, over a lifetime of more than 1000 million switching operations. Reed switches are used in telephone exchanges and door or window proximity switches in alarm systems. It also has industrial applications. Because the reeds and electrical contacts are encapsulated in a glass envelope, it can be used in explosive atmospheres.

Figure 1.10 Microswitches

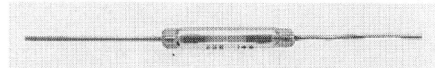

Figure 1.11 Reed switches

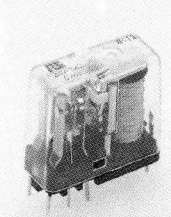

Figure 1.12 An electromagnetic relay

Electromagnetic relay

An electromagnetic relay is simply an electromagnet operating a number of switch contacts as shown in Figure 1.12. When a current is passed through the coil, the soft iron core becomes magnetised, attracts the iron armature and closes the switch contacts. The relay coil is electrically insulated from the switch contacts and, therefore, a relay is able to switch circuits operating at a different voltage to the coil operating voltage. The small current which energises the coil is also able to switch larger currents at the switch contacts. The switch part of the relay may have many poles controlling several circuits at once.

Miniature plug-in relays are popular in electronic circuits and intruder alarm circuits. However, all mechanical-electrical switches are limited in their speed of operation by the time taken physically for a movable contact to make or break a switch contact. Where extremely high-speed operations are required, the switching action must take place without physical movement. This is only possible using the properties of semiconductor materials in devices such as transistors and thyristors. They permit extremely high-speed switching without arcing and are considered in Chapter 6.

Overcurrent protection

Every piece of electronic equipment must incorporate some means of overcurrent protection. The term overcurrent can be sub-divided into *overload* current and *short circuit* current. An overload can be defined as a current which exceeds the rated value in an otherwise healthy circuit and a short circuit as an overcurrent resulting from a fault of negligible impedance between conductors. An overload may result in currents of two or three times the rated current flowing in the circuit, while short circuit currents may be hundreds of times greater than the rated current. In both cases the basic requirement for safety is that the fault current should be interrupted quickly and the circuit isolated from the supply. Fuses provide overcurrent protection when connected in the live conductor; they must not be connected in the neutral conductor. Circuit breakers may be used in place of fuses and the best protection of all is obtained when the equipment is connected to a residual current device. Figure 1.13 shows a cartridge fuse holder. Protection from excess current is covered in some detail in Chapter 8 of *Basic Electrical Installation Work*.

Supplies

The a.c. mains supply is probably the cheapest and most reliable source of electrical power. Low-voltage supplies are available from signal generators and power supply units which plug into the a.c. mains and these are considered in Chapter 5. However, there are occasions when the mains is unavailable, either because it has failed as in an emergency lighting scheme, or because the electronic equipment is to be used in a remote place. In these cases we must resort to portable battery power.

Figure 1.13 A fuse holder for cartridge fuses up to 15 A

Figure 1.14 Alkaline batteries

Batteries

The zinc-carbon

The zinc-carbon battery is the cheapest available battery. It has a zinc negative electrode, a manganese dioxide positive electrode and the electrolyte is a solution of ammonium chloride. They are available in the same range of sizes as the alkaline batteries shown in Figure 1.14, AAA, AA, C, D and PP3 having an emf. of 1.5 V. This is the most popular cell for low-current use such as torches. When the chemical reaction is exhausted the battery must be replaced.

The alkaline-manganese

The alkaline-manganese battery has a zinc negative electrode, a manganese dioxide positive electrode and the electrolyte is a strong solution of the alkali potassium hydroxide. This gives the battery up to four times the energy content of the standard zinc-carbon battery. They are available in a range of sizes as shown in Figure 1.14 and have an emf of 1.5 V. The battery is leak-proof being encased in a steel case and is ideal for use in calculators, personal radio cassettes, electric shavers and cameras. When the chemical reaction is exhausted the battery must be replaced.

The silver and mercury button cells

The silver and mercury button cells are small button-sized batteries with an emf respectively of 1.5 V and 1.35 V. They have a long life for such a small battery and are used where small occasional currents are required, such as in watches and cameras. Silver and mercury button cells are shown in Figure 1.15. When the chemical reaction is exhausted the battery must be replaced.

Batteries which must be replaced when the chemical reaction is exhausted are collectively called *primary cells*. A battery which can be repeatedly recharged is called a *secondary cell*, the most familiar of which is the lead-acid battery used in motor vehicles. *Lead-acid batteries* have lead plates immersed in a dilute solution of sulphuric acid. Each cell produces an emf of 2 V and most commonly six cells are grouped together in one 12 V battery. To maintain the battery in good condition the plates must always be covered and if necessary the electrolyte must be topped up with distilled water. The state of charge of a lead-acid battery can best be found by measuring the specific gravity of the electrolyte. When fully charged the specific gravity should have a value of 1.28 and when discharged or 'flat' a value of 1.15. Batteries should not be allowed to remain flat otherwise the lead sulphate hardens and cannot be changed back into lead dioxide and lead. They should be trickle charged to maintain a healthy state of readiness.

Figure 1.15 Silver and mercury button cells

Lead-acid batteries are used extensively on vehicles, as emergency supplies in public buildings and as a portable source of power in caravans. The advantage of a lead-acid battery is that it can be recharged when exhausted, but the major disadvantages are that they are very heavy and have a liquid electrolyte. The Nicad battery overcomes these difficulties and is suitable for electronic applications.

The Nickel-cadmium battery (Nicad)

The Nickel-cadmium battery (Nicad) has a nickel positive electrode, cadmium negative electrode and the electrolyte is potassium hydroxide. They are available in the same range of sizes as the Alkaline battery and are shown in Figure 1.16.

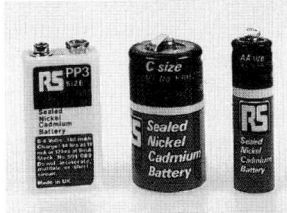

Figure 1.16 Nickel-cadmium rechargeable batteries

They have a high energy content but the terminal voltage is slightly lower than the equivalent size of Alkaline battery, 1.25 V for a Nicad and 1.5 V for the equivalent Alkaline. They can supply high currents, will operate successfully over a wide temperature range and are capable of accepting considerable overcharging when used with recommended chargers. Each battery will accept a minimum of 700 full discharge/charge cycles. They are used in the place of Alkaline batteries where a rechargeable capability is required, for example in self-contained emergency lighting luminaires.

Packaging electronic components

When we talk about packaging electronic components we are not referring to the parcel or box which contains the components for storage and delivery, but to the type of encapsulation in which the tiny semiconductor material is contained.

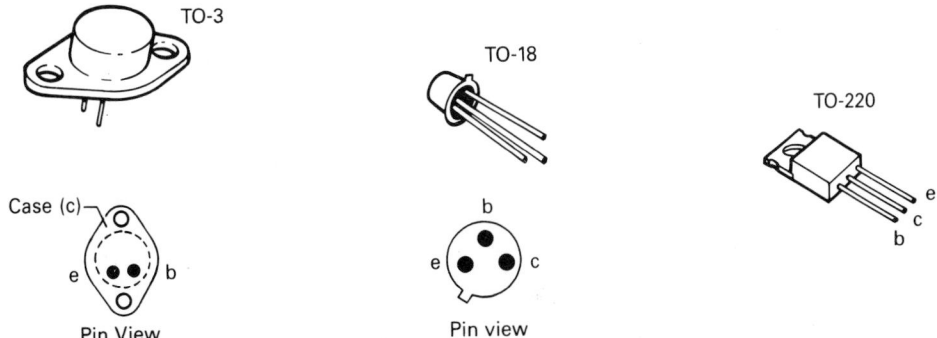

Figure 1.17 Three different package outlines for transistors

Obtaining information and components

Figure 1.17 shows three different package outlines for just one type of discrete component, the transistor. Identification of the pin connections for different packages is given within the text as each separate or discrete component is considered, particularly in Chapter 6, semiconductor devices. However, the Appendix aims to draw together all the information on pin connections and packages for easy reference.

Electricians use electrical wholesalers and suppliers to purchase electrical cable, equipment and accessories. Similar facilities are available in most towns and cities for the purchase of electronic components and equipment. There are also a number of national suppliers who employ representatives who will call at your workshop to offer technical advice and take your order. Some of these national companies also offer a 24-hour telephone order and mail order service. Their full-coloured, fully illustrated catalogue also contains an enormous amount of technical information. The names and addresses of these national companies are given in the Appendix. For local suppliers you must consult your local telephone directory and *Yellow Pages*. The Appendix of this book also contains some technical reference information.

CHAPTER 2

Electronic circuit assembly

To get a 'feel' for electronics you should take the opportunity to build some of the simple circuits described in Chapter 7 using the constructional methods described in this chapter. Practical electronics can be carried out with very few tools and little resources. A kitchen table, suitably protected, or a small corner of the electrical workshop is all that is required. The place chosen should be well lit, have a flat and dry area of about 1 m × 1 m and have access to a three-pin socket.

Working with others can also be a valuable source of inspiration and encouragement. Many technical colleges and evening institutes offer basic electronic courses which give someone new to electronics an opportunity to use the tools and equipment under guidance and at little cost. The City and Guilds of London offer a Supplementary Studies in Electronics examination which is particularly suitable for electricians who require a basic but formal qualification in electronics.

Safety precautions

Electricity can be dangerous. It can give a serious shock and it does cause fires. For maximum safety, the sockets being used for the electronic test and assembly should be supplied by a residual current circuit breaker. These sense fault currents as low as 30 mA so that a faulty circuit or piece of equipment can be isolated before the lethal limit to human beings of about 50 mA is reached. Plug-in RCCBs of the type shown in Figure 2.1 can now be bought very cheaply from any good electrical supplier or D.I.Y. outlet. All equipment should be earthed and fitted with a 2 A or 5 A fuse which is adequate for most electronics equipment. Larger fuses reduce the level of protection.

Another source of danger in electronic assembly is the hot soldering iron, which may cause burns or even start a fire. The soldering iron should always be placed in a soldering iron stand when not being used. The chances of causing a fire or burning yourself can be reduced by storing the soldering iron in its stand at the back of the workspace so that you don't have to lean over it when working.

So far we have been discussing the sensible safety precautions which everyone working with electricity should take. However, you or someone else in your workplace may receive an electric shock and I therefore offer the following guidance.

To a healthy person an electric shock from the mains is equivalent to a severe blow on the chest. It may make you jump back very suddenly and leave you breathless. Switch off the supply and sit

Figure 2.1 A plug in RCCB for safe electrical assembly

quietly while trying to discover why you received the shock. To the very young or very old, or someone with a less robust constitution, an electric shock from the mains can be serious. When this happens it is necessary to act quickly to prevent the electric shock becoming fatal.

Upon finding someone receiving an electric shock do not touch the person whilst they are still in contact with the electrical supply or you will risk being electrocuted yourself.

- switch off the supply or pull out the plug or
- remove the person from the supply without touching them e.g. push or pull them off with a broom or dry towel or coat
- if breathing or heartbeat has stopped, immediately apply resuscitation or cardiac massage until the patient recovers
- treat for shock
- call for medical assistance or dial 999.

First aid

Despite taking all sensible precautions, accidents do happen and *you* may be the only other person able to take action to assist someone in trouble. If you are not a qualified First Aider limit your help to obvious commonsense assistance and call for help *but* do remember that if someone's heart or breathing has stopped as a result of an accident, they have only minutes to live unless you act quickly.

Bleeding

If the wound is dirty, rinse it under clean running water. Clean the skin around the wound and apply a plaster, pulling the skin together.

If the bleeding is severe apply direct pressure to reduce the bleeding and raise the limb if possible. Apply a sterile dressing or pad and bandage firmly before obtaining professional advice.

To avoid possible contact with hepatitis or the AIDS virus when dealing with open wounds, First Aiders should avoid contact with fresh blood by wearing plastic or rubber protective gloves, or by allowing the casualty to apply pressure to the bleeding wound.

Burns

Remove heat from the burn to relieve the pain by placing the injured part under clean cold water.

Do not apply lotions or ointments. Do not break blisters or attempt to remove loose skin. Cover the injured area with a clean dry dressing.

Breathing stopped

Remove any restrictions from the face and any vomit, loose or false teeth from the mouth. Loosen tight clothing around the neck, chest and waist. To ensure a good airway, lay the casualty on his back and support the shoulders on some padding, tilt the head backwards and open the mouth. If the casualty is faintly breathing, lifting the tongue clear of the airway may be all that is necessary to restore normal breathing. However, if the casualty does not begin to breathe, open your mouth wide and take a deep breath, close the casualty's nose by pinching with your fingers, and, sealing your lips around his mouth, blow into his lungs until the chest rises. Remove your mouth and watch the casualty's chest fall. Continue this procedure at your natural breathing rate. If the mouth is damaged or you have difficulty making a seal around the casualty's mouth, close his mouth and inflate the lungs through his nostrils. Give artificial respiration until natural breathing is restored or until professional help arrives.

Heart stopped beating

This sometimes happens following a severe electric shock. The signs of cardiac arrest are: the casualty's lips may be blue, the pupils of his eyes widely dilated and the pulse in his neck cannot be felt. Act quickly and lay the casualty on his back. Kneel down beside him and place the heel of one hand in the centre of his chest. Cover this hand with your other hand and interlace the fingers. Straighten your arms and press down on his chest sharply with the heel of your hands and then release the pressure. Continue to do this 15 times at the rate of one push per second. Check the casualty's pulse; if none is felt, give two breaths of artificial respiration and then a further 15 chest compressions. Continue this procedure until the heartbeat is restored and the artificial respiration until normal breathing returns. Pay close attention to the condition of the casualty whilst giving heart massage. When a pulse is restored the blueness around the mouth will quickly go away and you should stop the heart massage. Look carefully at the rate of breathing. When this is also normal, stop giving

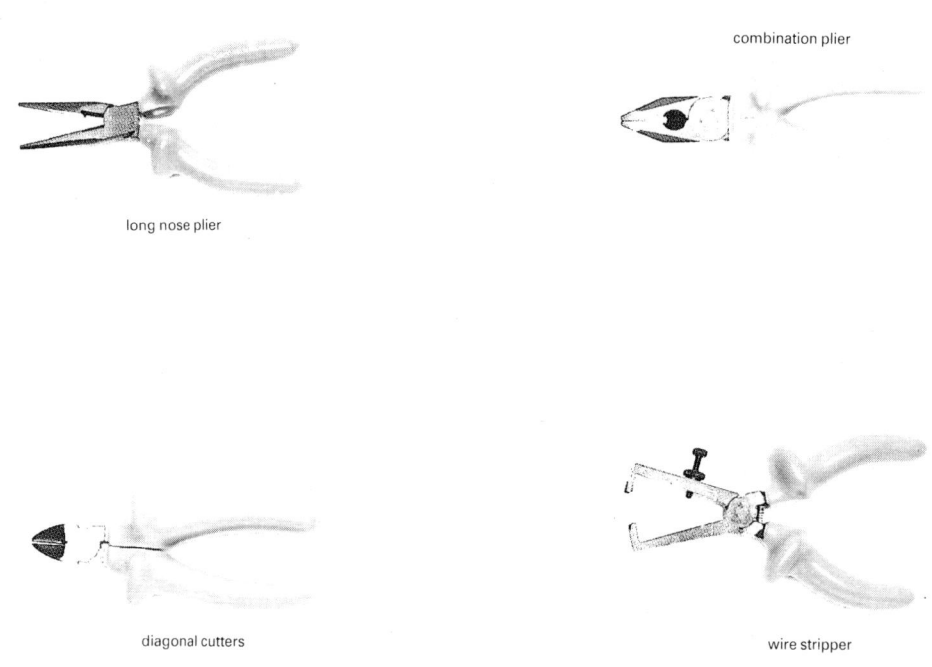

combination plier

long nose plier

diagonal cutters

wire stripper

Figure 2.2 Basic tools required for electronic assembly

artificial respiration. Treat the casualty for shock and obtain professional help.

Shock

Everyone suffers from shock following an accident. The severity of the shock depends upon the nature and extent of the injury. In cases of severe shock the casualty will become pale and his skin become clammy from sweating. He may feel faint, have blurred vision, feel sick and complain of thirst. Reassure the casualty that everything that needs to be done is being done. Loosen tight clothing and keep him warm and dry until help arrives. *Do not* move him unnecessarily or give him anything to drink.

Accident reports

Every accident should be reported to an employer and the details of the accident and treatment given entered in an 'Accident Book'. Failure to do so may influence the payment of compensation later.

Hand tools

Tools extend the physical capabilities of the human body. Good-quality, sharp tools are important to any craftsman. An electrician or electronic circuit assembler or repairer is no less a craftsman than a wood carver. Each must work with a high degree of skill and expertise and each must have

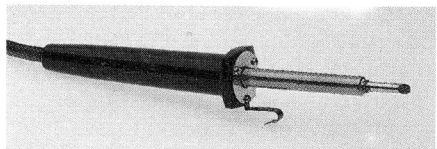

Figure 2.3 Electronic soldering iron and stand

sympathy and respect for the materials which they use. The basic tools required by anyone working with electrical equipment are those used to strip, cut and connect conductors and components. The tools required for successful electronic assembly are little different to those required for electrical installation work. The only possible additions being a pair of long-nosed pliers and a pair of wire strippers.

Figure 2.2 shows the basic hand tools required for successful electronic assembly.

Soldering irons

An electric soldering iron with the correct size bit is essential for making good-quality, permanent connections in electronic circuits. A soldering iron consists of a heat-insulated handle, supporting a heating element of between 15 W and 25 W. The bit is inserted into this element and heats up to a temperature of about 210°C by conduction. Various sizes of bits are available and are interchangeable.

Copper bits can be filed clean or rubbed with emery cloth until the tip is a bright copper colour. *Ironclad* bits must *not* be cleaned with a file or emery cloth but should be rubbed clean when they are hot, using a damp cloth or wet sponge.

Before the soldering iron can be used to make electrical connections, the bit must be '*tinned*' as follows:

– first clean the bit as described above
– plug in the soldering iron and allow it to heat up
– apply cored solder to the clean hot bit

– wipe off the excess solder with a damp cloth or damp sponge.

This will leave the soldering iron brightly 'tinned' and ready to be used. Figure 2.3 shows a 240 V general-purpose soldering iron and stand suitable for use in electronic assembly.

Soldering gun

Soldering guns of the type shown in Figure 2.4 are trigger-operated soldering irons. Within ten seconds of pressing the trigger the bit is at the working temperature of 315°C. The working temperature can be arrived at even more quickly with constant use. The plastic case holds a 240 V transformer having an isolated low-voltage high-current secondary circuit which is completed by the copper soldering bit. The bits are interchangeable and should be tinned and used in the same way as the general-purpose iron considered above.

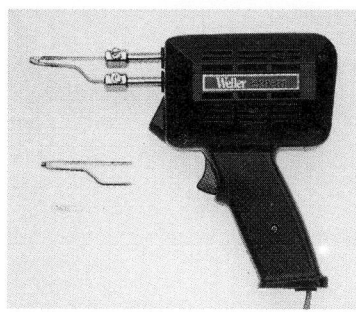

Figure 2.4 Instant soldering gun

Butane gas-powered soldering irons are also available. In appearance they are very similar to the general-purpose soldering iron shown in Figure 2.3, but without the mains cable. The handle acts as the fuel tank, various sizes of soldering bits are available and a protective cap is supplied to cover the hot end of the tool when not in use. The advantage of a gas soldering iron is that it can be used when a mains supply is not easily available.

The final choice of soldering iron will be influenced by many factors, frequency of use, where used, personal preference and cost. In 1990 the relevant costs were approximately £10 for the general-purpose iron, £20 for the soldering gun and £40 for the gas soldering iron.

Soldering

There are many ways of making suitable electrical connections and in electrical installation work a screwed terminal is the most common method. In electronics, the most common method of making permanent connections is by soldering the components into the circuit. Good soldering can only be achieved by effort and practice and you should, therefore, take the opportunity to practise the technique before committing your skills to the 'real' situation.

Soldering is an alloying process, whereby a small amount of soft metal (the solder) is made to run between the two metals to be joined, therefore, mixing or alloying them. Solder can be used to join practically any metals or alloys except those containing large amounts of chromium or aluminium, which must be welded or hard soldered.

Soft solders

Soft solders are so called because they are made up of the rather soft metals tin and lead in the proportion 40 to 60. Solders containing tin will adhere very firmly to most metals, providing that the surface of the metals to be joined is clean. Solder will not adhere to a tarnished or oxidised metal surface. This is because solder adheres by forming an alloy with the metal of the connection and this alloy cannot form if there is a film of oxide in the way.

Fluxes

Fluxes are slightly acid materials which dissolve an oxide film, leaving a perfectly clean surface to which the solder can firmly adhere. Rosin fluxes are the most suitable for electrical and electronic work. In electronics it is not convenient to apply the flux and solder separately, so they are combined as flux cored solder wire. This is solder wire with a number of cores of flux running the whole length of the wire. The multicore construction shown in Figure 2.5 ensures that there is always the correct proportion of flux for each soldered joint.

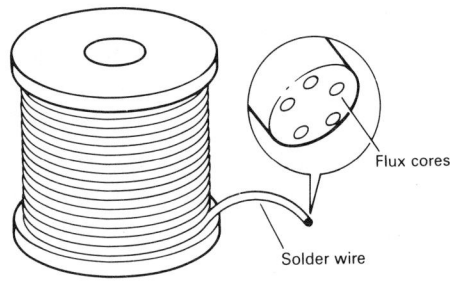

Figure 2.5 Construction of flux cored solder wire

Soldering techniques

As already mentioned, when soldering with an iron it is important to choose an iron with a suitable bit size. A 1.5 mm or 2.0 mm bit is suitable for most electronic connections but a 1.0 mm bit is better when soldering dual-in-line I.C. packages. The bit should be clean and freshly 'tinned'. The materials to be soldered must be free from grease and preferably pre-tinned. Electronic components should not need more cleaning than a wipe to remove dust or grease. The purpose of the soldering iron is to apply heat to the joint. If solder is first melted on to the bit, which is then used to transfer the solder to the joint, the active components of the flux will evaporate before the solder reaches the joint, and an imperfect or 'dry' joint will result. Also applying the iron directly to the joint oxidises the component surfaces, making them more difficult to solder effectively. The best method of making a 'good' soldered joint is to apply the cored solder to the joint and then melt the solder with the iron. This is the most efficient way of heating the termination, letting the solder and the flux carry

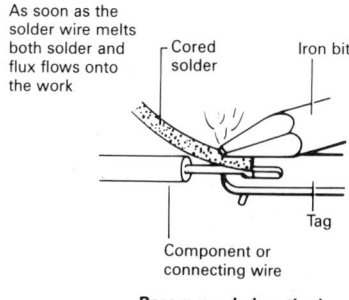

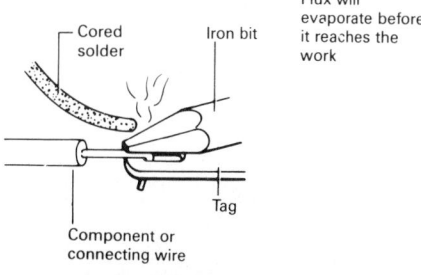

As soon as the solder wire melts both solder and flux flows onto the work — Cored solder   Iron bit   Tag   Component or connecting wire   **Recommended method**

Flux will evaporate before it reaches the work   — Cored solder   Iron bit   Tag   Component or connecting wire   **Incorrect method**

Figure 2.6 Soldering technique with multicore solders

the heat from the soldering bit on to the termination, as shown in Figure 2.6.

While the termination is heating up, the solder will appear dull, and then quite suddenly the solder will become bright and fluid, flowing around and 'wetting' the termination. Apply enough solder to cover the termination before removing the solder and then the iron. The joint should be soldered quickly. If attempts are made to improve the joint merely by continued heating and applying more flux and solder, the component or the cable insulation will become damaged by the heat and the connection will have excessive solder on it. The joint must not be moved or blown upon until the solder has solidified. A good soldered joint will appear smooth and bright, a bad connection or *dry joint* will appear dull and the solder may be in a 'blob' or appear spiky.

Dry joints

Dry joints may occur because the components or termination are dirty or oxidised, or because the soldering temperature was too low, or too little flux was used. Dry joints do not always make an

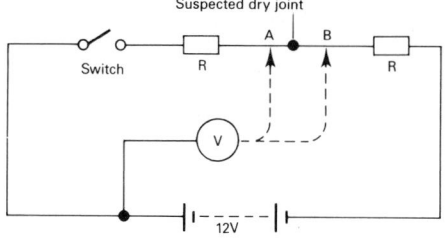

Figure 2.7 Testing a suspected dry joint

electrical connection or the connection has a high resistance which deteriorates with time and may cause trouble days or weeks later. A suspected dry joint can be tested as shown in Figure 2.7. If the joint is 'dry' the voltmeter will read 12 V at position B, just to the right and 0 V at position A to the left of the joint. If the joint is found to be dry, the connection must be remade.

Component assembly and soldering

Soft solder is not as strong as other metals and, therefore, the electronic components must be shaped at the connection site to give extra strength. This can be done by bending the connecting wires so that they hook together or by making the joint area large. Special *lead forming* or *wire shaping* tools are available which both cut and shape the components connecting wires ready for soldering. Figure 2.8 shows a suitable tag terminal connection, Figure 2.9, a suitable pin terminal connection and Figure 2.10 a suitable stripboard connection.

All wires must be cut to length before assembly because it is often difficult to trim them after soldering. Also the strain of cutting after soldering may weaken the joint and encourage dry joints. If the wires must be cut after soldering, cutters with a shearing action should be used as shown in Figure 2.11. Side cutters have a pinching action and the shock of the final pinch-through, identified by a sharp click, may fracture the soldered joint or damage the component.

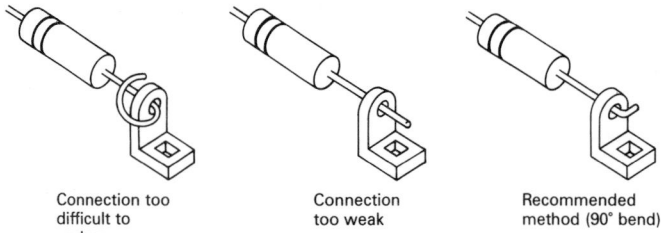

Figure 2.8 Shaping conductors to give strength to electrical connections to tag terminals

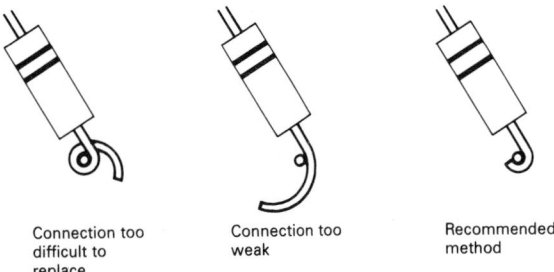

Figure 2.9 Shaping conductors to give strength to electrical connections to pin terminals (plan view)

Most electronic components are very sensitive and are easily damaged by excess heat. Soldered joints must not, therefore, be made close to the body of the component or the heat transferred from the joint may cause some damage. When components are being soldered into a circuit the heat from the soldering iron at the connection must be diverted or 'shunted' away from the body of the component. This can be done by placing a pair of long-nosed pliers or a crocodile clip between the soldered joint and the body of the component as shown in Figure 2.12.

Components such as resistors, capacitors and transistors are usually cylindrical, rectangular or disc shaped with round wire terminations. They should be shaped, mounted and soldered into the circuit as previously described and shown in Figure 2.13. A small clearance should be left between the body of the component and the circuit board, to allow convection currents to circulate, which encourages cooling. The vertical mounting method permits many more components to be mounted on the circuit board but the horizontal method gives better mechanical support to the component.

Desoldering

If it is necessary to replace an electronic component, the old, faulty component must first be removed from the circuit board. To do this the

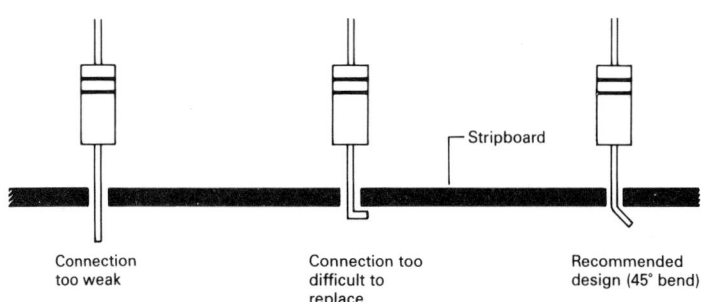

Figure 2.10 Shaping conductors to give strength to electrical connections to stripboard

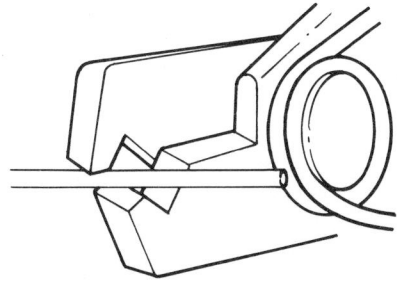

(a) Shearing action

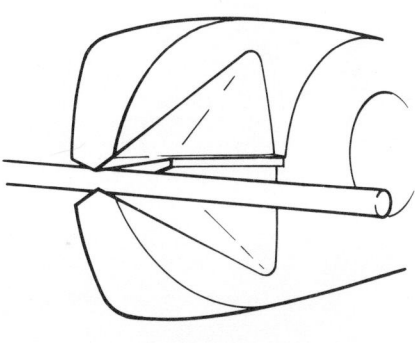

(b) Pinching action

Figure 2.11 Wire cutting

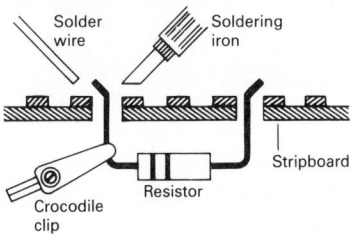

Figure 2.12 Using a crocodile clip as a heat shunt

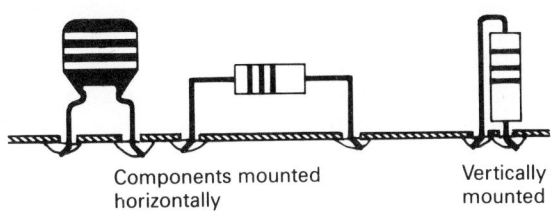

Figure 2.13 Vertical and horizontal mounting of components

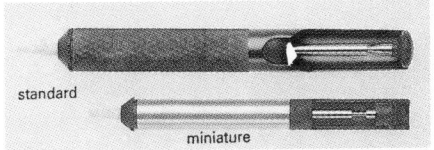

Figure 2.14 A desoldering tool

solder of the old joint is first liquefied by applying a hot iron to the joint. The molten solder is then removed from the joint with a desoldering tool. The desoldering tool works like a bicycle pump in reverse and is shown in Figure 2.14. The tool is made ready by compressing the piston down on to a latch position which holds it closed. The nozzle is then placed into the pool of molten solder and the latch release button pressed. This releases the plunger which shoots out, sucking the molten solder away from the joint and into the body of the desoldering tool.

Removing faulty transistors

First identify the base, collector and emitter connections so that the new component can be correctly connected into the circuit. Remove the solder from each leg with the soldering iron and desoldering tool before removing the faulty transistor. Then, with the aid of a pair of long-nosed pliers, pull the legs of the transistor out of the circuit board. An alternative method is to cut the three legs with a pair of side cutters before desoldering and then remove the individual legs with a pair of long-nosed pliers.

Removing faulty Integrated Circuits (chips)

Remove the solder from each leg of the IC with the soldering iron and desoldering tool and then pull the IC clear of the circuit board. If it has been firmly established that the IC is faulty, it may be removed from the circuit board by cutting the body from the connecting pins before desoldering and removing the individual pins with a pair of long-nosed pliers.

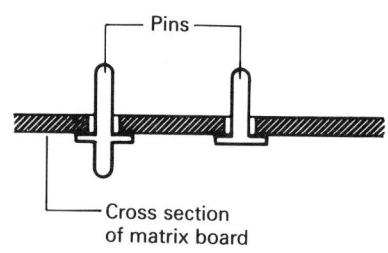

Cross section
of matrix board

Figure 2.15 Matrix board and double-sided and single-sided pin inserts

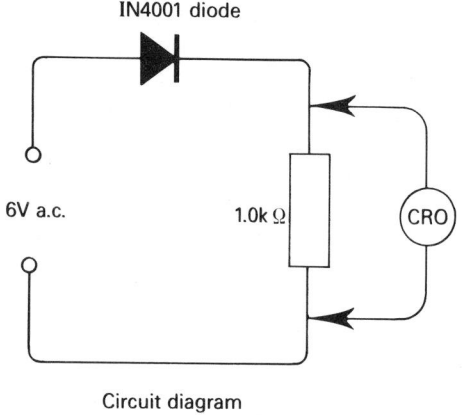

Circuit diagram

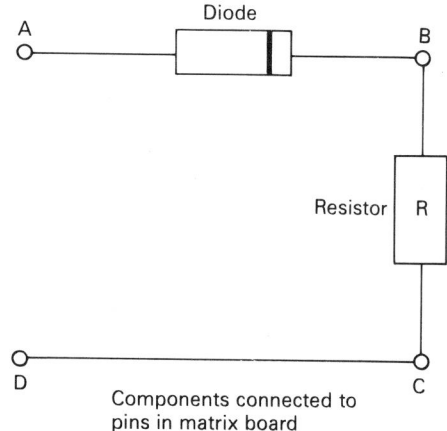

Components connected to
pins in matrix board

Figure 2.16 Circuit diagram converted to a component layout

Circuit boards

Permanent circuits require that various discrete components be soldered together on some type of insulated board. Three types of board can be used, matrix, strip and printed circuit board, the base material being plastic synthetic resin bonded paper (SRBP).

Matrix boards

Matrix boards have a matrix of holes on 0.1 inch centres as shown in Figure 2.15. Boards are available in various sizes, the 149 × 114 mm board is pierced with 58 × 42 holes and the 104 × 65 mm board has 39 × 25 holes. Matrix pins press into any of the holes in the board and provide a terminal post to which components and connecting wires can be soldered. Single sided or double sided matrix pins are available. Double sided pins have the advantage that connections can be made on the underside of the board as well as on the top. The hole spacing of 0.1 inch makes the board compatible with many electronic components. Plug-in relays, d.i.l. integrated circuits and many sockets and connectors all use 0.1 inch spacing at their connections.

Matrix board is probably the easiest and cheapest way to build simple electronic circuits. It is recommended that inexperienced circuit builders construct the circuit on the matrix board using a layout which is very similar to the circuit diagram to reduce the possibility of mistakes.

Suppose, for example, that we intend to build the very simple circuit shown in Figure 2.16. First we would insert four pins into the matrix board as shown. The diode would then be connected between pins A and B, taking care that the anode was connected to pin A. The resistor would be connected between pins B and C and a wire linked between pins C and D. The a.c. supply from a signal generator would be connected to pins A and D by 'flying' leads and the oscilloscope leads to pins B and C. This circuit would show half wave rectification. When planning the conversion of circuit diagrams into a matrix board layout it helps to have a positional reference system so that we know where to push the pins in the matrix board.

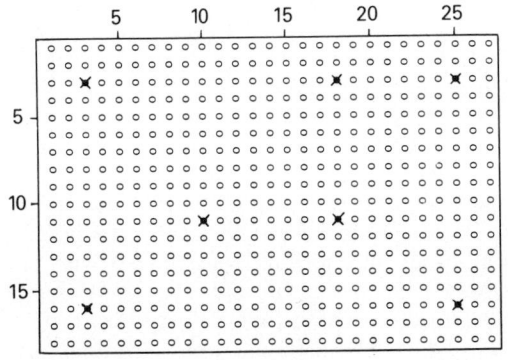

Figure 2.17 Matrix board pin reference system

The positional reference system

The positional reference system used with matrix boards uses a simple grid reference system to identify holes on the board. This is achieved by counting along the columns at the top of the board, starting from the left and then counting down the

rows. For example, the position reference point 4 : 3 would be 4 holes from the left and 3 holes down. The board should be prepared as follows:

- turn the matrix board so that a manufactured straight edge is to the top and left-hand side
- use a felt tip pen to mark the holes in groups of five along the top edge and down the left-hand edge as shown in Figure 2.17.

The pins can then be inserted as required. Figure 2.15 shows a number of pin reference points. Counting from the left-hand side of the board there are 3 : 3, 3 : 16, 10: 11, 18 : 3, 18 : 11, 25 : 3 and 25 : 16.

Stripboard or Veroboard

Stripboard or Veroboard is a matrix board with continuous copper strip attached to one side by adhesive. The copper strips link together rows of holes so that connections can be made between components inserted into holes on a particular row as shown in Figure 2.18. The components are assembled on the plain board side with the component leads inserted through the holes and soldered in place to the copper strip.

The copper strips are continuous but they can be broken using a strip cutter or small drill. The drill or cutter is placed on the hole where the break is to be made and then rotated a few turns between the fingers until the very thin copper strip is removed leaving a circle as shown in Figure 2.18.

Stripboard is very useful because the copper strips take the place of the wire links required with plain matrix boards. Components can easily be mounted vertically on stripboard which leads to high-density small area circuits being assembled. It

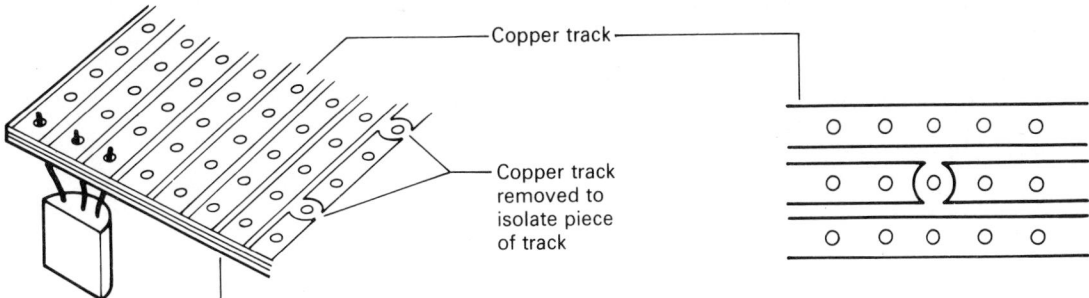

Figure 2.18 Stripboard or veriboard

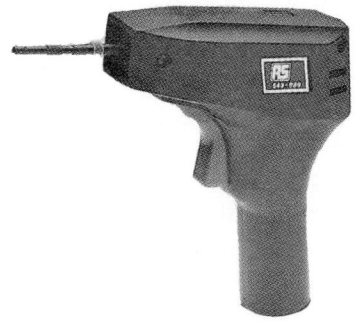

battery operated

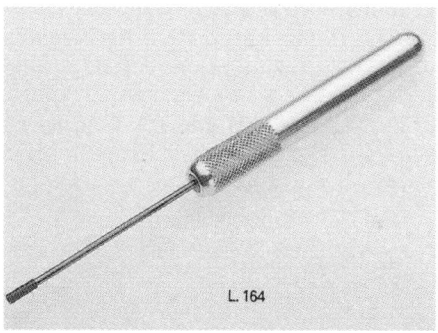

L. 164

hand operated

Figure 2.19 Wire wrapping tools

is, however, more expensive than matrix board because of the additional cost of the copper, most of which is not used in the circuit. Also some translation of the circuit diagram is required before it can be assembled on the stripboard. Excessive heat from the soldering iron can melt the adhesive and cause the copper to peel from the insulating board. Heat should only be applied long enough to melt the solder and secure the component.

Printed circuit boards (PCB)

Printed circuit boards (PCB) are produced by chemically etching a copper-clad epoxy glass board so that a copper pattern is engraved on one side of the board. The pattern provides the copper conducting paths making connections to the various components of the circuit. After etching, small holes are drilled for the components which are inserted from the plain side of the board and soldered on to the copper conductor. The copper pattern replaces the lengths of wire used to connect components on to the matrix board.

The copper foil is very thin and is attached to the board with an adhesive. Excessive heat from the soldering iron can melt the adhesive and cause the track to peel from the insulating board. The board should not be flexed, otherwise hairline cracks may appear but go unnoticed until intermittent faults occur in the circuit later.

The process of designing and making a PCB is quite simple but does require specialised equipment and for that reason we will not consider it further here.

Wire wrapping

As the electronic industry increasingly uses advanced technology, the demand for a faster, more reliable and inexpensive method of making electrical connections has increased. Electronic equipment today has lots of components with many terminals in a very small space. To make electrical connections to these high-density electronic circuits, the electronic industry has developed a new solderless wire wrapping technique.

A wire wrapped connection is made by winding a special insulated wire of 30 AWG (0.25 mm) around the sharp corners of a square pin inserted into the circuit board. The winding tension causes the corners of the pin to cut into the wire, producing a good electrical connection which will not unwind and is as good as or better than a soldered joint. This method of connection was developed by the Bell Telephone Laboratories of the Western Electric Company.

Pneumatic or electric tools are preferred for production work but battery- or hand-operated tools as shown in Figure 2.19 are used for service and repair work.

To make the connection

–   the end of the wire is inserted into the bit of the tool and then bent back at 90° to the tool.
–   the bit is placed over the terminal pin
–   the tool is rotated clockwise a few turns without undue pressure.
–   the bit is removed from the terminal pin.
–   the connection is made.

One advantage of wire wrapping is the ease with

which a wire can be removed from a terminal because of an error or wiring modification. An unwrap tool is placed over the terminal pin, rotated anticlockwise, and the connection is removed in seconds without damage to the terminal pin.

Wire wrapping is a precision technique and the bit size and wire diameter must be compatible with the terminal pin size if a good electrical connection is to be made.

Breadboards or prototype

Breadboarding is the name given to solderless temporary circuit building by pressing wires and component leads into holes in the prototype board. This method is used for building temporary circuits for testing or investigation.

The S-DeC

The S-DeC prototype board is designed for interconnecting discrete components. The hole spacing and hole connections do not suit d.i.l. I.C. packages. Each board has 70 phosphor-bronze contact points arranged in two sections, each of which has 7 parallel rows of 5 connected contact points. The case is formed in high impact polystyrene and the individual boards may be interlocked to create a larger working area. The S-DeC can be supplied with a vertical bracket for mounting switches or variable resistors as shown in Figure 2.20.

The professional

The professional prototype board is designed for the interconnection of many different types of component. The hole spacing of 0.1 inch allows d.i.l. I.C. packages to be plugged directly into the board. Each board has 47 rows of 5 interconnected nickel-silver contacts each side of a central channel and a continuous row at the top and bottom which may be used as power supply rails. The case is formed in high impact thermoplastic and the individual boards may be interlocked to create a larger working area. A vertical side bracket is also supplied for mounting switches or variable resistors as shown in Figure 2.21.

Interconnection methods

A plug and socket provide an ideal method of connecting or isolating components and equip-

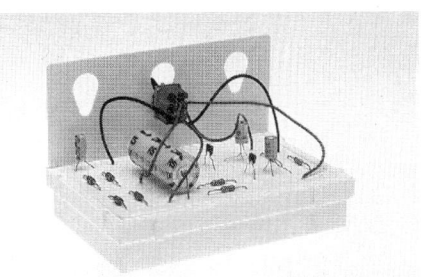

Figure 2.20 S-DeC prototype board used for temporary circuit building

Figure 2.21 Professional prototype board used for temporary circuit building

ment which cannot be permanently connected. In electrical installation work we usually need to make plug and socket connections between three conductors on single phase circuits and five conductors on three phase systems. In electronics we often need to make multiple connections between circuit boards or equipment. However, the same principles apply, that is the plug and socket must be capable of separation, but while connected they must make a good electrical contact. Also the plug and socket must incorporate some method of preventing reverse connection.

P.c.b. edge connectors

A range of connectors are available which make direct contact to printed circuit boards as shown in Figure 2.22. Multiple connectors are available

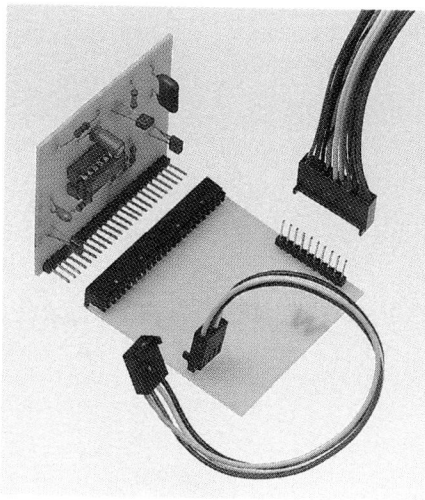

Figure 2.22 PCB edge connectors

with a contact pitch of 0.1 inch so that they can be soldered into circuit boards. The plug and socket can then be used as edge connectors to make board-to-board and cable-to-board connections.

Ribbon cable connectors

A ribbon cable is a multicore cable laid out as flat strip or ribbon strip. A range of connectors is made to connect ribbon cable, as shown in Figure 2.23, which is used for making board-to-board interconnections and to connect computer peripherals such as VDUs and printers.

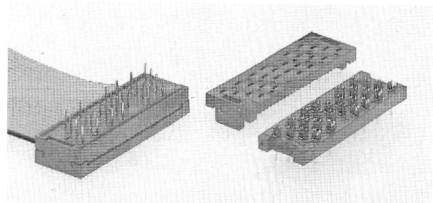

Figure 2.23 Ribbon cable connector

DIN connectors

DIN-style audio connectors are available for making up to 8 connections as shown in Figure 2.24 and used when frequent connection and disconnection is required between a small number of contacts.

Jack connectors

Jack connectors are used when frequent connections are to be made between two or three poles on, for example, headphones or microphones. They are available in three sizes, sub-miniature (2.5 mm), miniature (3.5 mm) and commercial (0.25 inch) as shown in Figure 2.25.

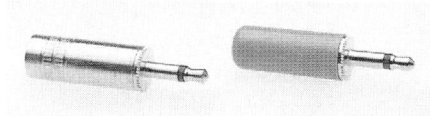

Figure 2.25 Jack connectors

All the above connectors may be terminated onto the cable end by a soldered, crimped or cable displacement method. When a soldered connection is to be made the cable end must be stripped of its insulation, tinned and then terminated. A crimped connection also requires that the insulation be removed and the prepared cable end inserted into a lug, which is then crimped using an appropriate tool. The insulation displacement method of connection is much quicker to make because the cable ends do not require stripping or preparing. The connection is made by pressing insulation piercing tines or prongs into the cable which displaces the insulation to make an electrical

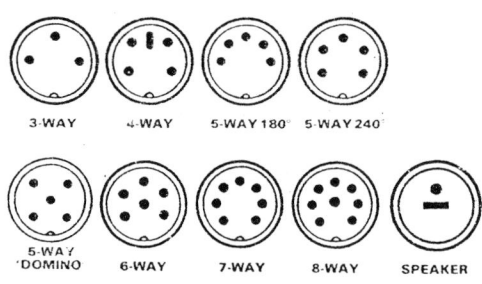

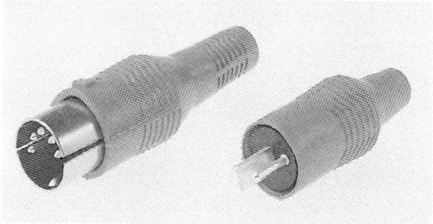

Figure 2.24 'DIN'-style audio connectors

connection with the conductor. This method is used extensively when terminating ribbon cable and for making rapid connections to the existing wiring system of motor vehicles.

Fault finding

The best way to avoid problems in electronic circuit assembly is to be always alert while working, to think about what you are doing and always try to be neat. If, despite your best efforts, the circuit does not work as it should when tested, then follow a logical test procedure which will usually find the fault in the shortest possible time. First carry out a series of visual tests.

1. Is the battery or supply correctly connected?
2. Is the battery flat or the supply switched on?
3. Is the circuit constructed *exactly* as it should be according to the circuit diagram?
4. Are all the components in place?
5. Check the values of all the components.
6. Are all the components such as diodes, capacitors, transistors and I.C.s connected the correct way round?
7. Have all connections and links been made?
8. Have all the necessary breaks been made in the stripboard, e.g. between the I.C.?
9. Are all the soldered joints good?
10. Are any of the components hot or burnt?

If the fault has not been identified by the first ten tests, ask someone else to carry them out. You may have missed something which will be obvious to someone else. If the visual tests have failed to identify the fault, then further meter tests are called for.

11. Check the input voltage and the output voltage. Check the mid-point voltage between components which are connected in series with the supply.
12. Variable resistors may suffer from mechanical wear. Check the voltage at the wiper as well as across the potentiometer.
13. Check the coil voltage on relays; if this is low, the coil contacts may not be making.
14. Is the diode connected correctly? Short circuit the diode momentarily with a wire link to see if the circuit works. If it does the diode is open circuit.
15. Check resistor capacitor circuits by momentarily shorting out the capacitor and then observing the charging voltage. If it does not charge, the resistor may be open circuit. If it charges instantly, the resistor may be short circuit. Check the polarity of electrolytic capacitors. Check the capacitor leads for breaks where the lead enters the capacitor body.
16. Check the base-emitter voltage of the transistor. A satisfactory reading would be between 0.6 V and 1.0 V. Temporarily connect a 1 kΩ resistor between the positive supply and the base connection. If the transistor works, the base feed is faulty. If it does not work, the transistor is faulty.
17. Short out the anode and cathode of the thyristor. If the load operates, the thyristor or the gate pulse is faulty. If the load does not operate, the load is faulty.

The testing of capacitors and resistors is further discussed in Chapter 1, while the testing of discrete semiconductor components is dealt with in Chapter 6.

CHAPTER 3

Circuit theory

This chapter brings together most of the circuit theory in the training courses which lead to 'approved electrician' status for electricians. Electricians should, therefore, consider this chapter as revision while readers from other disciplines might like to 'dip into' the theory on a 'need to know' basis.

Units

Very early units of measurement were based on the things easily available – length of a stride, the distance from the nose to the outstretched hand, the weight of a stone and the time-lapse of one day. Over the years, new units were introduced and old ones were modified. Different branches of science and engineering were working in isolation, using their own units, and the result was an overwhelming variety of units.

In all branches of science and engineering there is a need for a practical system of units which everyone can use. In 1960 the General Conference of Weights and Measures agreed to an international system called the Système International d'Unités (abbreviated to SI units). SI units are based upon a small number of fundamental units from which all other units may be derived, see Table 3.1.

Like all metric systems, SI units have the advantage that prefixes representing various multiples or submultiples may be used to increase or decrease the size of the unit by various powers of ten. Some of the more common prefixes and their symbols are shown in Table 3.2.

Basic circuit theory

All matter is made up of atoms which arrange themselves in a regular framework within a material. The atom is made up of a central, positively

Table 3.1   S.I. units

SI Unit	Measure of	Symbol
The fundamental units		
Metre	Length	m
Kilogram	Mass	kg
Second	Time	s
Ampere	Electric current	A
Kelvin	Thermodynamic temperature	K
Candela	Luminous intensity	cd
Some derived units		
Coulomb	Charge	C
Joule	Energy	J
Newton	Force	N
Ohm	Resistance	Ω
Volt	Potential difference	V
Watt	Power	W

Table 3.2   Prefixes for use with S.I. units

Prefix	Symbol	Multiplication factor		
mega	M	$\times 10^6$	or	$\times 1\,000\,000$
kilo	k	$\times 10^3$	or	$\times 1\,000$
hecto	h	$\times 10^2$	or	$\times 100$
deca	da	$\times 10$	or	$\times 10$
deci	d	$\times 10^{-1}$	or	$\div 10$
centi	c	$\times 10^{-2}$	or	$\div 100$
milli	m	$\times 10^{-3}$	or	$\div 1\,000$
micro	μ	$\times 10^{-6}$	or	$\div 1\,000\,000$

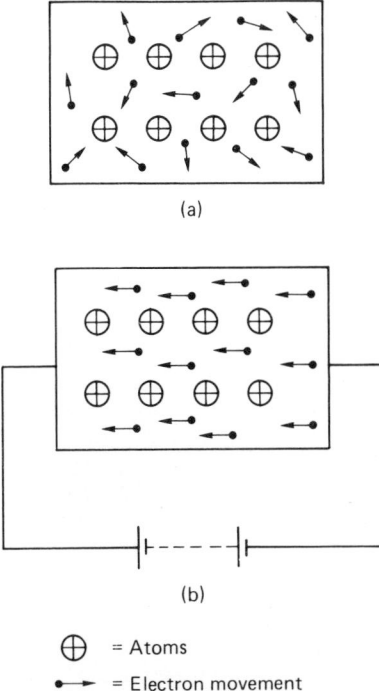

(a)

(b)

$\oplus$ = Atoms

$\bullet\!\!-\!\!\rightarrow$ = Electron movement

Figure 3.1 Atoms and electrons on a material

charged nucleus, surrounded by negatively charged electrons. The electrical properties of a material depend largely upon how tightly these electrons are bound to the central nucleus.

A *conductor* is a material in which the electrons are loosely bound to the central nucleus and are, therefore, free to drift around the material at random from one atom to another as shown in Figure 3.1(a). Materials which are good conductors include copper, brass, aluminium and silver.

An *insulator* is a material in which the outer electrons are tightly bound to the nucleus and so there are no free electrons to move around the material. Good insulating materials are PVC, rubber, glass and wood.

If a battery is attached to a conductor as shown in Figure 3.1(b), the free electrons drift purposefully in one direction only. The free electrons close to the positive plate of the battery are attracted to it, since unlike charges attract, and the free electrons near the negative plate will be repelled from it. For each electron entering the positive terminal of the battery, one will be ejected from the negative terminal, so the number of electrons in the conductor remains constant.

This drift of electrons within a conductor is known as an electric *current*, measured in amperes and given the symbol I. For a current to continue to flow, there must be a complete circuit for the electrons to move around. If the circuit is broken by opening a switch, for example, the electron flow, and therefore the current, will stop immediately.

To cause a current to flow continuously around a circuit, a driving force is required, just as a circulating pump is required to drive water round a central heating system. This driving force is the *electromotive force* (abbreviated to emf) and is the energy which causes the current to flow in a circuit. Each time an electron passes through the source of emf, more energy is provided to send it on its way around the circuit.

An emf is always associated with energy conversion, such as chemical to electrical in batteries and mechanical to electrical in generators. The energy introduced into the circuit by the emf is transferred to the load terminals by the circuit conductors. The *potential difference* (abbreviated to p.d.) is the change in energy levels measured across the load terminals. This is also called the volt drop or terminal voltage, since emf and p.d. are both measured in volts. Every circuit offers some opposition to current flow which we call the circuit *resistance* measured in ohms, to commemorate the famous experimenter George Simon Ohm, who was responsible for the analysis of electrical circuits. The symbol Ω represents an ohm.

In 1826, Ohm published details of an experiment he had made to investigate the relationship between the current passing through and the potential difference between the ends of wire. As a result of this experiment he arrived at the following law, known as *Ohm's law*: 'The current passing through a conductor under constant temperature conditions is proportional to the potential difference across the conductor'.

$$V = I \times R \text{ (V)}$$

Transposing this formula, we also have

$$I = \frac{V}{R} \text{(A)} \quad \text{and} \quad R = \frac{V}{I} \text{(}\Omega\text{)}$$

Example 1

An electric heater when connected to a 240 V supply was found to take a current of 4 A. Calculate the element resistance.

$$R = \frac{V}{I} \ (\Omega)$$

$$\therefore \quad R = \frac{240 \text{ V}}{4 \text{ A}} = 60 \ \Omega$$

Example 2

The insulation resistance measured between phase conductors on a 415 V supply was found to be 2 MΩ. Calculate the leakage current.

$$I = \frac{V}{R} \ (\text{A})$$

$$\therefore \quad I = \frac{415}{2 \times 10^6 \ \Omega} = 207.5 \times 10^{-6} \text{ A} = 207.5 \ \mu\text{A}.$$

Example 3

When a 4 Ω resistor was connected across the terminals of an unknown d.c. supply, a current of 3 A flowed. Calculate the supply voltage.

$$V = I \times R \ (\text{V})$$

$$\therefore \quad V = 3 \text{ A} \times 4 \ \Omega = 12 \text{ V}.$$

Resistivity (symbol ρ – the Greek letter 'rho')

The resistance or opposition to current flow varies for different materials, each having a particular constant value. If we know the resistance of say one metre of a material, then the resistance of five metres will be five times the resistance of one metre.

The *resistivity* of a material is defined as the resistance of a sample of unit length and unit cross-section. Typical values are given in Table 3.3. Using the constants for a particular material we can calculate the resistance of any length and thickness of that material from the equation.

$$R = \frac{\rho l}{a} \ (\Omega)$$

where ρ = the resistivity constant for the material (Ωm)
l = the length of the material (m)
a = the cross-sectional area of the material (m²).

Table 3.3 gives the resistivity of silver as 16.4×10^{-9} ohm metre which means that a sample of silver one metre long and one metre in cross-section will have a resistance of 16.4×10^{-9} ohm.

Table 3.3   Resistivity values

Material	Resistivity [ohm metre]
Silver	16.4×10^{-9}
Copper	17.5×10^{-9}
Aluminium	28.5×10^{-9}
Brass	75.0×10^{-9}
Iron	100.0×10^{-9}

Example 1

Calculate the resistance of 100 metres of copper cable of 1.5 mm² cross-sectional area if the resistivity of copper is taken as 17.5×10^{-9} Ωm.

$$R = \frac{\rho l}{a} \ (\Omega)$$

$$R = \frac{17.5 \times 10^{-9} \ \Omega\text{m} \times 100 \text{ m}}{1.5 \times 10^{-6} \ \text{m}^2} = 1.16 \ \Omega.$$

Example 2

Calculate the resistance of 100 metres of aluminium cable of 1.5 mm² cross-sectional area if the resistivity of aluminium is taken as 28.5×10^{-9} Ωm.

$$R = \frac{\rho l}{a} \ (\Omega)$$

$$R = \frac{28.5 \times 10^{-9} \ \Omega\text{m} \times 100 \text{ m}}{1.5 \times 10^{-6} \ \text{m}^2} = 1.9 \ \Omega.$$

The above examples show that the resistance of an aluminium cable is some 60% greater than a copper conductor of the same length and cross-section. Therefore, if an aluminium cable is to replace a copper cable, the conductor size must be increased to carry the rated current as given by the tables in Appendix 9 of the IEE Regulations.

Electrostatics

If a battery is connected between two insulated plates, the emf of the battery forces electrons from one plate to another until the p.d. between the plates is equal to the battery emf.

The electrons flowing through the battery constitute a current, I, amperes which flow for a time, t, seconds. The plates are then said to be charged. The amount of charge transferred

$$Q = It \text{ (C)}$$

Figure 3.2 shows the charges on a capacitor's plates.

When the voltage is removed the charge Q is trapped on the plates, but if the plates are joined together, the same quantity of electricity, $Q = It$, will flow back from one plate to the other, so discharging them. The property of two plates to store an electric charge is called its *capacitance*.

By definition, a capacitor has a capacitance of one farad when a p.d. of one volt maintains a charge of one coulomb on that capacitor, or

$$C = \frac{Q}{V} \text{ (F)}.$$

Collecting these important formulae together we have

$$Q = It = CV.$$

Capacitors

A capacitor consists of two metal plates, separated by an insulating layer called the dielectric. It has the ability of storing a quantity of electricity as an excess of electrons on one plate and a deficiency on the other.

Example

A 100 μF capacitor is charged by a steady current of 2 mA flowing for 5 seconds. Calculate the total charge stored by the capacitor and the p.d. between the plates.

$$Q = It \text{ (C)}$$

$$Q = 2 \times 10^{-3} \text{ A} \times 5 \text{ s} = 10 \text{ mC}$$

$$Q = CV$$

$$V = \frac{Q}{C} \text{ (V)}$$

$$V = \frac{10 \times 10^{-3} \text{ C}}{100 \times 10^{-6} \text{ F}} = 100 \text{ V}.$$

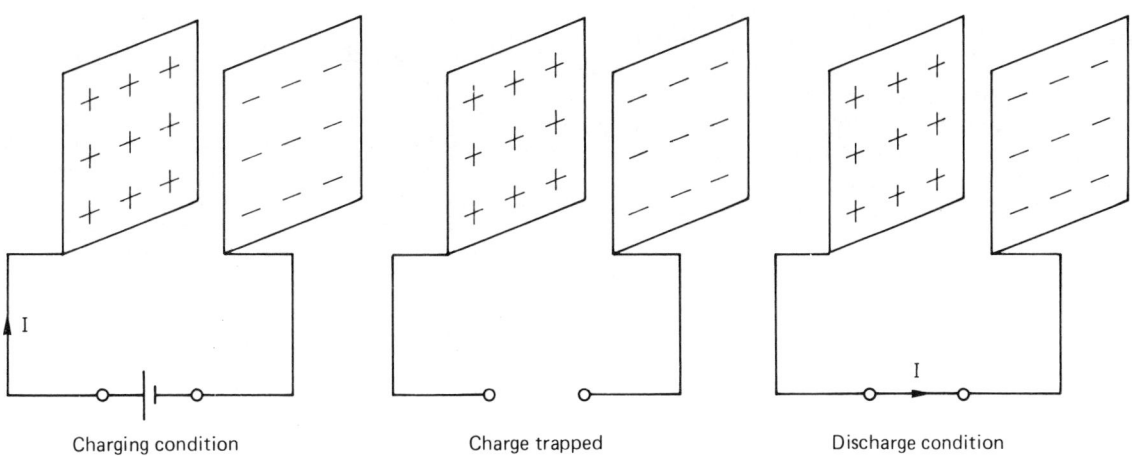

Charging condition   Charge trapped   Discharge condition

Figure 3.2 The charge on a capacitor's plates

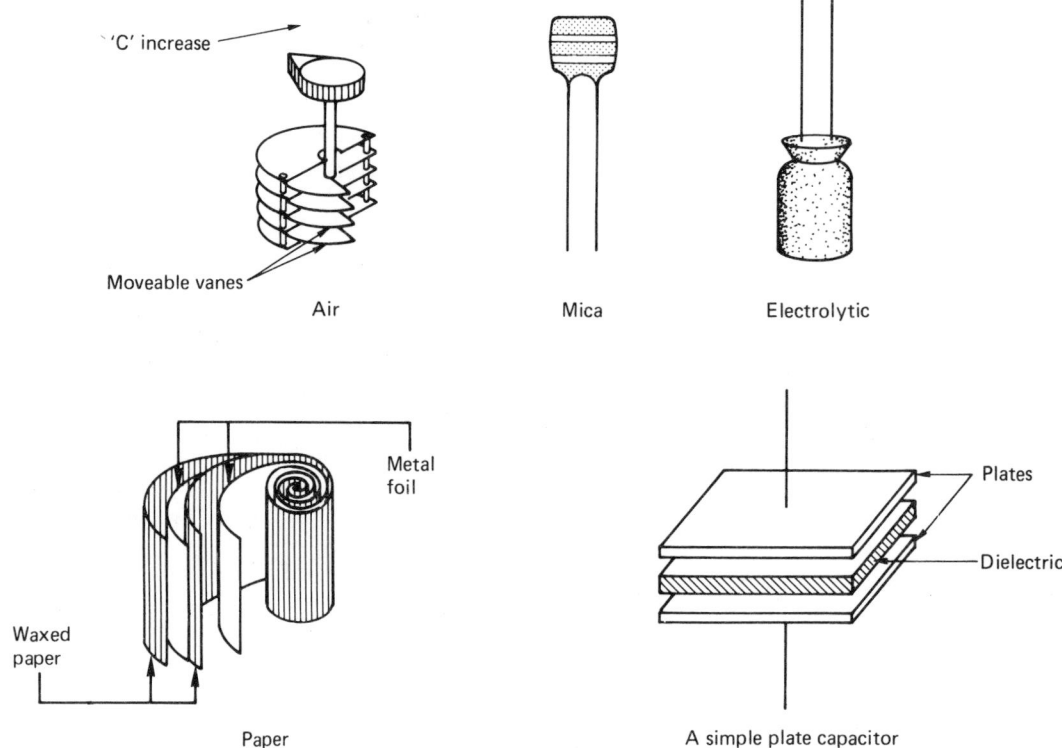

'C' increase

Moveable vanes

Air

Mica

Electrolytic

Metal foil

Waxed paper

Paper

Plates

Dielectric

A simple plate capacitor

Figure 3.3 Construction and appearance of capacitors

The p.d. which may be maintained across the plates of a capacitor is determined by the type and thickness of the dielectric medium. Capacitor manufacturers usually indicate the maximum safe working voltage for their products.

Capacitors are classified by the type of dielectric material used in their construction. Chapter 1 shows the general construction and appearance of capacitors to be found in electronic work.

Charging capacitors

Connecting a voltage to the plates of a capacitor causes it to charge up to the potential of the supply. This involves electrons moving around the circuit to create the necessary charge conditions and, therefore, this action does not occur instantly, but takes some time. This scientific fact has many applications in electronic circuits.

C-R circuits

Figure 3.4 shows the circuit diagram for a simple C-R circuit and the graphs drawn from the meter readings. It can be seen that:

(a)   initially the current has a maximum value and decreases slowly to zero as the capacitor charges and
(b)   initially the capacitor voltage rises rapidly but then slows down, increasing gradually until the capacitor voltage is equal to the supply voltage when fully charged.

The mathematical name for the shape of these curves is an *exponential* curve and, therefore, we say that the capacitor voltage is growing exponentially while the current is decaying exponentially during the charging period. The *rate* at which the capacitor charges is dependent upon the *size* of the

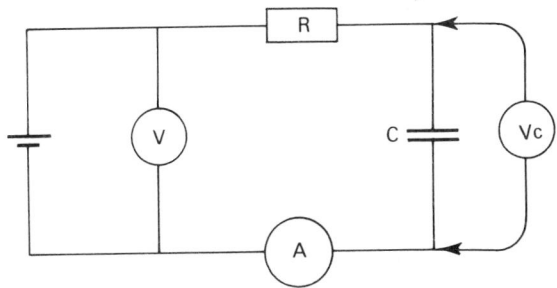

Figure 3.4 A C-R circuit

capacitor and resistor. The bigger the values of C and R, the longer will it take to charge the capacitor. The time taken to charge a capacitor by a *constant* current is given by the *Time Constant* of the circuit which is expressed mathematically as

$T = CR$ (s) where T is the time in seconds.

Example 1

A 60 μF capacitor is connected in series with a 20 kΩ resistor across a 12 V supply. Determine the time constant of this circuit.

$T = CR$ (s)

$\therefore T = 60 \times 10^{-6} \, \text{F} \times 20 \times 10^3 \, \Omega.$

$T = 1.2$ (s)

We have already seen in practice the capacitor is not charged by a *constant* current but, in fact, charges exponentially. However, it can be shown by experiment that in *one* time constant the capacitor will have reached about 63% of its final steady value, taking about five times the time constant to become fully charged. Therefore, in 1.2(s) the 60 μF capacitor of Example 1 will have reached about 63% of 12 V and after 5 T, that is six seconds, will be fully charged at 12 V.

Resistors

In an electrical or electronic circuit, resistors may be connected in series, in parallel or in various combinations of series and parallel connections.

Series-connected resistors

In any series circuit a current I will flow through all parts of the circuit as a result of the potential difference supplied by the battery V_T. Therefore, we say that in a series circuit the current is common throughout that circuit.

When the current flows through each resistor in the circuit, R_1, R_2, and R_3, as, for example, in Figure 3.5 there will be a voltage drop across that resistor whose value will be determined by the values of I and R, since from Ohm's Law $V = I \times R$. The sum of the individual voltage drops, V_1, V_2, and V_3, will be equal to the total voltage V_T.

We can summarise these statements as follows: For any series circuit

I = common throughout the circuit

$V_T = V_1 + V_2 + V_3$

Let us call this Equation 1.

Let us call the total circuit resistance R_T.

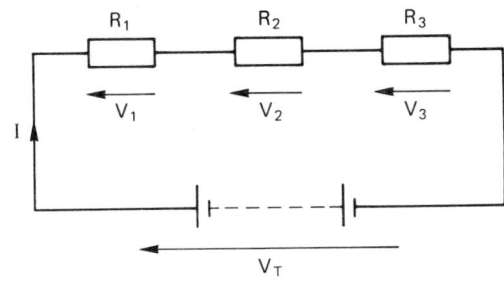

Figure 3.5 A series circuit

From Ohm's law we know that $V = I \times R$ and therefore

The total voltage $\quad\quad V_T = I \times R_T$
The voltage drop across R_1 is $\quad V_1 = I \times R_1$
The voltage drop across R_2 is $\quad V_2 = I \times R_2$
The voltage drop across R_3 is $\quad V_3 = I \times R_3$

Let us call these Equations 2.

We are looking for an expression for the total resistance in any series circuit and, if we substitute Equations 2 into Equation 1 we have:

$$V_T = V_1 + V_2 + V_3$$
$$\therefore I \times R_T = I \times R_1 + I \times R_2 + I \times R_3$$

Now, since I is common to all terms in the equation, we can divide both sides of the equation by I. This will cancel out I to leave us with an expression for the circuit resistance:

$$R_T = R_1 + R_2 + R_3$$

Note   The derivation of this formula is given for information only. Craft students need only state the expression: $R_T = R_1 + R_2 + R_3$ for series connections.

Parallel-connected resistors

In any parallel circuit, as shown in Figure 3.6, the same voltage acts across all branches of the circuit. The total current will divide when it reaches a resistor junction, part of it flowing in each resistor. The sum of the individual currents I_1, I_2 and I_3 for example in Figure 3.6 will be equal to the total current I_T. We can summarise these statements as follows:

For any parallel circuit

V = common to all branches of the circuit

$$I_T = I_1 + I_2 + I_3$$

Let us call this Equation 1.

Let us call the total resistance R_T.

From Ohm's law we know that $I = \dfrac{V}{R}$

and therefore

The total current $\quad\quad I_T = \dfrac{V}{R_T}$

The current through R_1 is $\quad I_1 = \dfrac{V}{R_1}$

The current through R_2 is $\quad I_2 = \dfrac{V}{R_2}$

The current through R_3 is $\quad I_3 = \dfrac{V}{R_3}$

Let us call these Equations 2.

We are looking for an expression for the equivalent resistance R_T in any parallel circuit and, if we substitute Equation 2 into Equation 1 we have:

$$I_T = I_1 + I_2 + I_3$$
$$\therefore \frac{V}{R_T} = \frac{V}{R_1} + \frac{V}{R_2} + \frac{V}{R_3}$$

Now since V is common to all terms in the equation, we can divide both sides by V, leaving us with an expression for the circuit resistance:

$$\frac{1}{R_T} = \frac{1}{R_1} + \frac{1}{R_2} + \frac{1}{R_3}$$

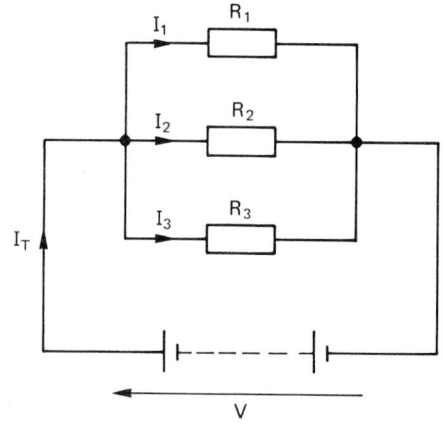

Figure 3.6 A parallel circuit

Note The derivation of this formula is given for information only. Craft students need only state the expression: $1/R_T = 1/R_1 + 1/R_2 + 1/R_3$ for parallel connections.

Example 1

Three $6\,\Omega$ resistors are connected (a) in series, see Figure 3.7 and (b) in parallel, see Figure 3.8, across a 12V battery. For each method of connection, find the total resistance and the values of all currents and voltages.

For any series connection

$$R_T = R_1 + R_2 + R_3$$
$$\therefore R_T = 6\,\Omega + 6\,\Omega + 6\,\Omega = 18\,\Omega$$

Total current $I_T = \dfrac{V}{R_T}$

$$\therefore I_T = \frac{12\,\text{V}}{18\,\Omega} = 0.66\,\text{A}$$

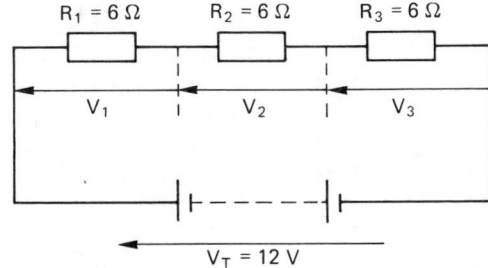

Figure 3.7 Resistors in series

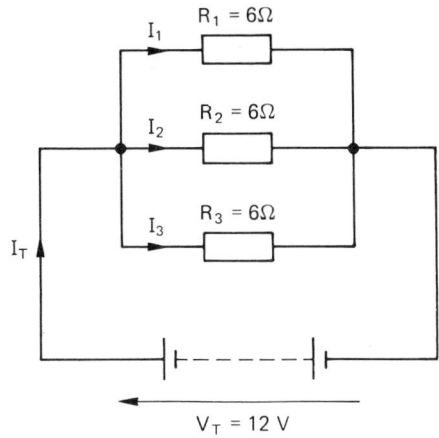

Figure 3.8 Resistors in parallel

The voltage drop across R_1 is

$$V_1 = I \times R_1$$
$$\therefore V_1 = 0.66\,\text{A} \times 6\,\Omega = 4\,\text{V}$$

The voltage drop across R_2 is

$$V_2 = I \times R_2$$
$$\therefore V_2 = 0.66\,\text{A} \times 6\,\Omega = 4\,\text{V}$$

The voltage drop across R_3 is

$$V_3 = I \times R_3$$
$$\therefore V_3 = 0.66\,\text{A} \times 6\,\Omega = 4\,\text{V}.$$

For any parallel connection,

$$\frac{1}{R_T} = \frac{1}{R_1} + \frac{1}{R_2} + \frac{1}{R_3}$$

$$\therefore \frac{1}{R_T} = \frac{1}{6\,\Omega} + \frac{1}{6\,\Omega} + \frac{1}{6\,\Omega}$$

$$\frac{1}{R_T} = \frac{1+1+1}{6\,\Omega} = \frac{3}{6\,\Omega}$$

$$R_T = \frac{6\,\Omega}{3} = 2\,\Omega$$

Total current $I_T = \dfrac{V}{R_T}$

$$\therefore I_T = \frac{12\,\text{V}}{2\,\Omega} = 6\,\text{A}$$

The current flowing through R_1 is

$$I_1 = \frac{V}{R_1}$$

$$\therefore I_1 = \frac{12\,\text{V}}{6\,\Omega} = 2\,\text{A}$$

The current flowing through R_2 is

$$I_2 = \frac{V}{R_2}$$

$$\therefore I_2 = \frac{12\,\text{V}}{6\,\Omega} = 2\,\text{A}$$

The current flowing through R_3 is

$$I_3 = \frac{V}{R_3}$$

$$\therefore I_3 = \frac{12\ V}{6\ \Omega} = 2\ A$$

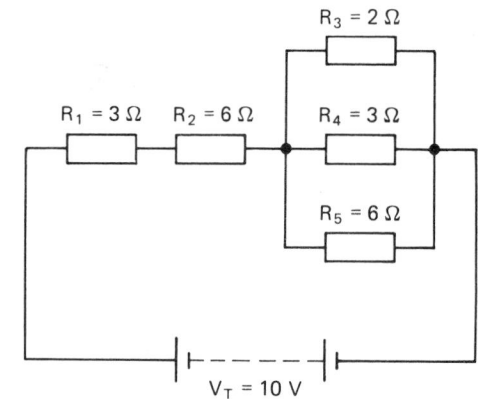

Figure 3.9 A series/parallel circuit

Series and parallel combinations

The most complex arrangement of series and parallel resistors can be simplified into a single equivalent resistor by combining the separate rules for series and parallel resistors.

Example 1

Resolve the circuit shown in Figure 3.9 into a single resistor and calculate the potential difference across each resistor. By inspection the circuit contains a parallel group R_3, R_4, R_5, and a series group consisting of R_1 and R_2 in series with the equivalent resistor for the parallel branch. Consider the parallel group R_3, R_4, R_5, we will label this group R_P.

$$\frac{1}{R_P} = \frac{1}{R_3} + \frac{1}{R_4} + \frac{1}{R_5}$$

$$\frac{1}{R_P} = \frac{1}{2\ \Omega} + \frac{1}{3\ \Omega} + \frac{1}{6\ \Omega}$$

$$\frac{1}{R_P} = \frac{3+2+1}{6\ \Omega} = \frac{6}{6\ \Omega}$$

$$R_P = \frac{6\ \Omega}{6} = 1\ \Omega$$

Figure 3.9 may now be represented by the more simple equivalent shown in Figure 3.10. Since all resistors are now in series

$$R_T = R_1 + R_2 + R_3$$

$$\therefore R_T = 3\ \Omega + 6\ \Omega + 1\ \Omega = 10\ \Omega.$$

Thus, the circuit may be represented by a single equivalent resistor of value $10\ \Omega$ as shown in

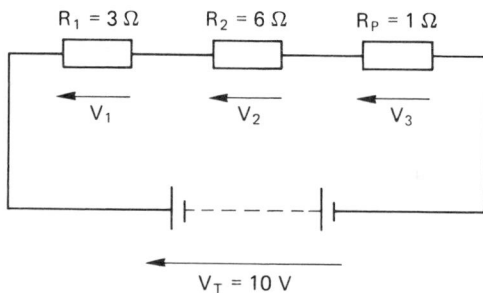

Figure 3.10 Equivalent series circuit

Figure 3.11. The total current flowing in the circuit may be found by using Ohm's law $I = V/R$

$$I_T = \frac{V_T}{R_T} = \frac{10\ V}{10\ \Omega} = 1\ A$$

The potential difference across the individual resistors are:

$$V_1 = 1 \times R_1 = 1\ A \times 3\ \Omega = 3\ V$$
$$V_2 = 1 \times R_2 = 1\ A \times 6\ \Omega = 6\ V$$
$$V_P = 1 \times R_P = 1\ A \times 1\ \Omega = 1\ V$$

Since the same voltage acts across all branches of a parallel circuit the same p.d. of one volt will exist across each resistor in the parallel branch R_3, R_4 and R_5. Six volts will be dropped across R_2 and three volts across R_1.

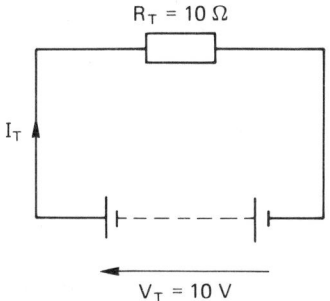

Figure 3.11 Single equivalent resistor for Figure 3.9

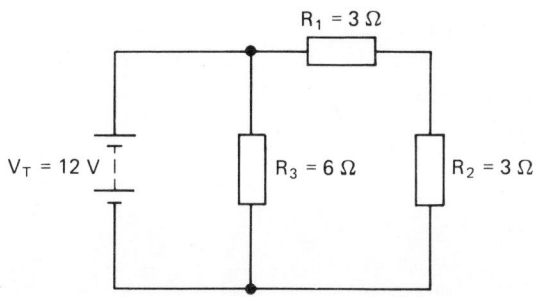

Figure 3.12 A series/parallel circuit for Example 2

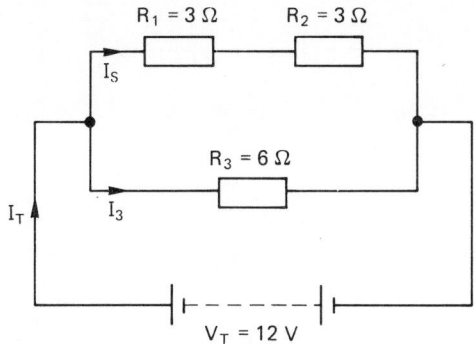

Figure 3.13 Equivalent circuit for Example 2

Example 2

Determine the total resistance and the current flowing through each resistor for the circuit shown in Figure 3.12. By inspection, it can be seen that R_1 and R_2 are connected in series while R_3 is connected in parallel across R_1 and R_2. The circuit may

be more easily understood if we redraw it as in Figure 3.13. For the series branch, the equivalent resistor can be found from

$$R_s = R_1 + R_2$$
$$\therefore R_s = 3\,\Omega + 3\,\Omega = 6\,\Omega$$

Figure 3.13 may now be represented by a more simple equivalent circuit as shown in Figure 3.14. Since the resistors are now in parallel, the equivalent resistance may be found from

$$\frac{1}{R_T} = \frac{1}{R_s} + \frac{1}{R_3}$$

$$\therefore \frac{1}{R_T} = \frac{1}{6\,\Omega} + \frac{1}{6\,\Omega}$$

$$\frac{1}{R_T} = \frac{1+1}{6\,\Omega} = \frac{2}{6\,\Omega}$$

$$R_T = \frac{6\,\Omega}{2} = 3\,\Omega$$

The total current is

$$I_T = \frac{V}{R_T} = \frac{12\,\text{V}}{3\,\Omega} = 4\,\text{A}$$

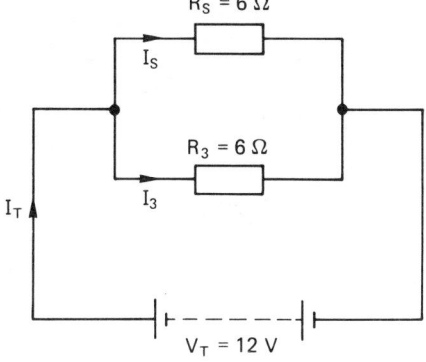

Figure 3.14 Simplified equivalent circuit for Example 2

Let us call the current flowing through resistor R_3 I_3:

$$\therefore I_3 = \frac{V}{R_3} = \frac{12\,\text{V}}{6\,\Omega} = 2\,\text{A}$$

Let us call the current flowing through both resistors R_1 and R_2, as shown in Figure 3.13, I_s:

$$\therefore I_s = \frac{V}{R_s} = \frac{12\,\text{V}}{6\,\Omega} = 2\,\text{A}$$

Capacitors in combination

Capacitors, like resistors, may be joined together in various combinations of series or parallel connections, see Figure 3.15. The equivalent capacitance C_T, of a number of capacitors is found by the application of similar formulae to that used for resistors and discussed earlier in this chapter.

Note   The form of the formulae is the opposite way round to that used for series and parallel

resistors. The most complex arrangement of capacitors may be simplified into a single equivalent capacitor by applying the separate rules for series or parallel capacitors in a similar way to the simplification of resistive circuits.

Example 1

Capacitors of 10 μF and 20 μF are connected first in series and then in parallel, as shown in Figures 3.16 and 3.17. Calculate the effective capacitance for each connection. Consider the series connection:

$$\frac{1}{C_T} = \frac{1}{C_1} + \frac{1}{C_2}$$

$$\frac{1}{C_T} = \frac{1}{10\,\mu\text{F}} + \frac{1}{20\,\mu\text{F}}$$

$$\frac{1}{C_T} = \frac{2+1}{20\,\mu\text{F}} = \frac{3}{20\,\mu\text{F}}$$

$$\therefore C_T = \frac{20\,\mu\text{F}}{3} = 6.66\,\mu\text{F}$$

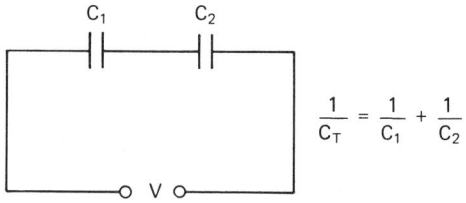

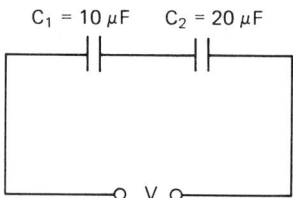

Figure 3.16 Series capacitors

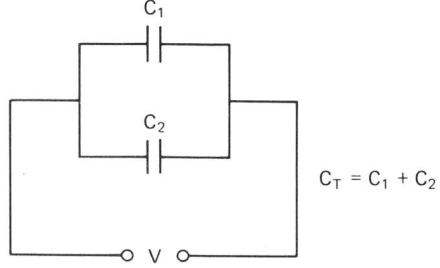

Figure 3.15 Connection and formulae of series or parallel capacitors

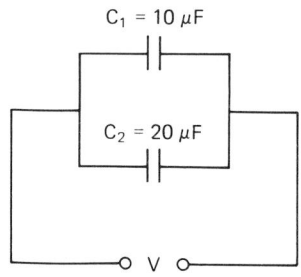

Figure 3.17 Parallel capacitors

Consider the parallel connection:

$$C_T = C_1 + C_2$$

$$\therefore C_T = 10\,\mu F + 20\,\mu F = 30\,\mu F$$

Therefore, when capacitors of 10 μF and 20 μF are connected in series their combined effect is equivalent to a single capacitor of 6.66 μF, but when the same capacitors are connected in parallel their combined effect is equal to a capacitor of 30 μF.

Power and energy

Power

Power is the rate of doing work and is measured in watts.

$$\text{Power} = \frac{\text{Work done}}{\text{Time taken}}\,(W)$$

In an electrical circuit

Power = Voltage × current (W)

Let us call this Equation 1.

Now, from Ohm's law

Voltage = $I \times R$ (V)

Let us call this Equation 2.

$$\text{Current} = \frac{V}{R}\,(A)$$

Let us call this Equation 3.

Substituting Equation 2 into Equation 1 we have,

Power = $(I \times R) \times \text{Current} = I^2 \times R$ (W)

and substituting Equation 3 into Equation 1 we have

$$\text{Power} = \text{Volts} \times \frac{V}{R} = \frac{V^2}{R}\,(W)$$

We can find the power of a circuit by using any of the three formulae

$$P = V \times I,\ P = I^2 \times R,\ P = \frac{V^2}{R}$$

Energy

Energy is a concept which engineers and scientists use to describe the ability to do work in a circuit or system.

Energy = Power × Time

but, since Power = Voltage × Current

then Energy = Voltage × Current × Time

The SI unit of energy is the Joule, where time is measured in seconds. For practical electrical installation circuits this unit is very small and, therefore, the kilowatt hour (kWh) is used for domestic and commercial installations. Electricity Board meters measure 'units' of electrical energy where each 'unit' is one kWh.

Energy in joules = Voltage × Current × Time in seconds.

Energy in kWh = kW × Time in hours.

Example 1

A domestic immersion heater is switched on for 40 minutes and takes 12.5 A from a 240 V supply. Calculate the energy used during this time.

Power = Voltage × Current
Power = 240 V × 12.5 A = 3000 W or 3 kW
Energy = kW × Time in hours

$$\text{Energy} = 3\,kW \times \frac{40\ min}{60\ min/h} = 2\,kWh.$$

This immersion heater uses 2 kWh in 40 minutes or 2 'units' of electrical energy every 40 minutes.

Example 2

An electronic Hi-Fi system takes a current of 200 mA when connected to the 240 V mains. Calculate (a) the power, (b) the energy in Joules and (c) the energy in kWh consumed by this system per hour.

For (a) Power = Voltage × Current
        Power = 240 V × 200 × 10^{-3} A = 48 W.
For (b) Energy = Power × Time in seconds
        Energy = 48 W × 60 × 60 s = 172800 J.
For (c) Energy = kW × Time in hours
        Energy = 0.048 × 1 = 0.048 kWh.

In general, electronic equipment consumes much less energy than electrical installation appliances. You can also see from the answer to part (b) above that the Joule is a very small unit of energy and, therefore, not a practical unit for most electrical purposes.

Example 3

Two 50 Ω resistors may be connected to a 200 V supply. Determine the power dissipated by the resistors when they are connected (a) in series, (b) each resistor separately connected and (c) in parallel.

For (a) the equivalent resistance when resistors are connected in series is given by

$$R_T = R_1 + R_2 \, \Omega$$

$$\therefore R_T = 50 \, \Omega + 50 \, \Omega = 100 \, \Omega$$

$$\text{Power} = \frac{V^2}{R_T} \, (\text{W})$$

$$\therefore \text{Power} = \frac{200 \, \text{V} \times 200 \, \text{V}}{100 \, \Omega} = 400 \, \text{W}$$

For (b) each resistor separately connected has a resistance of 50 Ω

$$\text{Power} = \frac{V^2}{R} \, (\text{W})$$

$$\therefore \text{Power} = \frac{200 \, \text{V} \times 200 \, \text{V}}{50 \, \Omega} = 800 \, \text{W}.$$

For (c) the equivalent resistance when resistors are connected in parallel is given by

$$\frac{1}{R_T} = \frac{1}{R_1} + \frac{1}{R_2}$$

$$\therefore \frac{1}{R_T} = \frac{1}{50 \, \Omega} + \frac{1}{50 \, \Omega}$$

$$\frac{1}{R_T} = \frac{1+1}{50 \, \Omega} = \frac{2}{50 \, \Omega}$$

$$R_T = \frac{50 \, \Omega}{2} = 25 \, \Omega$$

$$\text{Power} = \frac{V^2}{R_T} \, (\text{W})$$

$$\therefore \text{Power} = \frac{200 \, \text{V} \times 200 \, \text{V}}{25 \, \Omega} = 1,600 \, \text{W}$$

This example shows that by connecting 50 Ω resistors together in different combinations of series or parallel connections, we can obtain various power outputs; in this example 400, 800 and 1,600 watts.

Instrument connections

Electrical and electronic measuring instruments must be chosen and connected into the circuit to be tested with great care for the reasons described in Chapter 5.

Ammeters are connected in series with the load and voltmeters in parallel across the load. Wattmeters contain both current and voltage coils within the instrument. Since Watts = Volts × Amps, the voltage coil must be connected in parallel and the current coil in series with the load as shown in Figure 3.18.

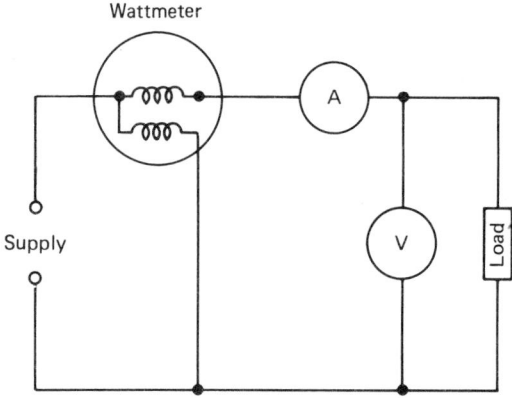

Figure 3.18 Wattmeter, Ammeter and Voltmeter connected to a load

Alternating current theory

The supply which we obtain from a car battery is a uni-directional or d.c. supply, whereas the mains electricity supply is alternating or a.c. as shown in Figure 3.19.

Most electrical equipment makes use of alternating current supplies and for this reason a knowledge of alternating waveforms and their effect upon resistive, capacitive and inductive loads is necessary for all practising electricians.

When a coil of wire is rotated inside a magnetic field a voltage is induced in the coil. The induced voltage follows the mathematical law known as a sinusoidal law and, therefore, we can say that a sine wave has been generated. Such a waveform has the characteristics displayed in Figure 3.20.

In the UK we generate electricity at a frequency of 50 Hz and the time taken to complete each cycle is given by

$$T = \frac{1}{f} \text{ (s)}$$

$$\therefore T = \frac{1}{50} \text{ Hz} = 0.02 \text{ s}$$

An alternating waveform is constantly changing from zero to a maximum, first in one direction,

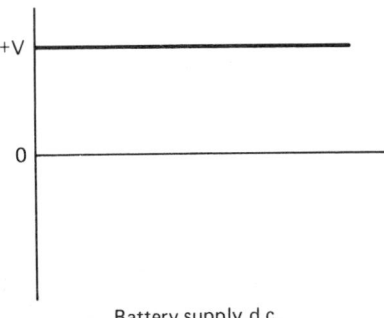

Battery supply d.c.

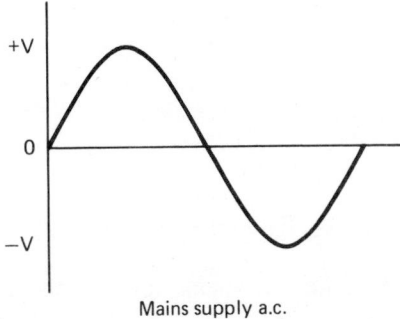

Mains supply a.c.

Figure 3.19 Uni-directional and alternating supply

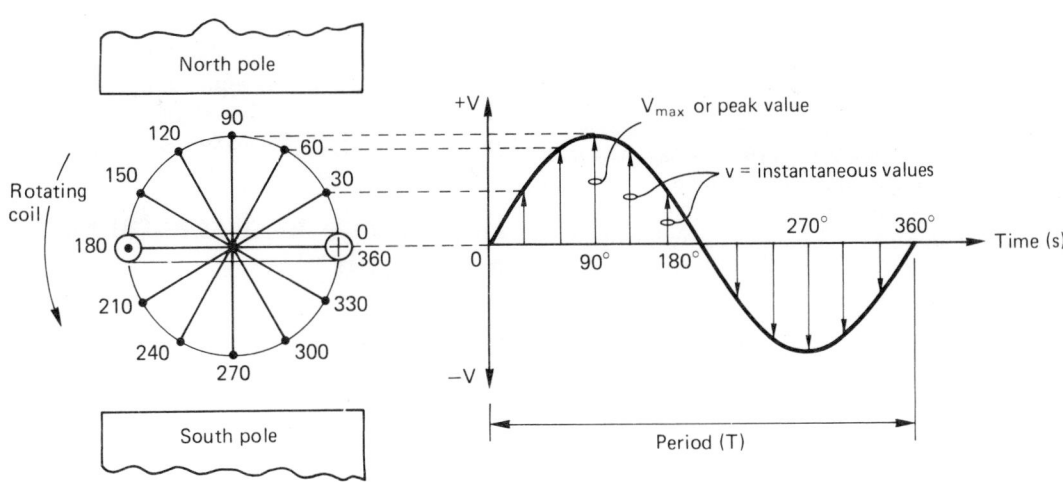

Figure 3.20 Characteristics of a sine wave

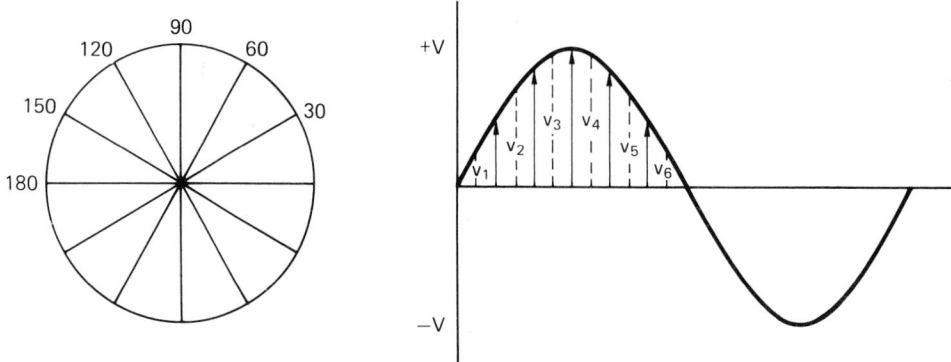

Figure 3.21 Sinusoidal waveform showing instantaneous values of voltage

then in the opposite direction and so the instantaneous values of the generated voltage are always changing. A useful description of the electrical effects of an a.c. waveform can be given by the maximum, average and rms values of the waveform.

The maximum or peak value is the greatest instantaneous value reached by the generated waveform. Cable and equipment insulation levels must be equal to or greater than this value.

The average value is the average over one half-cycle of the instantaneous values as they change from zero to a maximum and can be found from the following formulae applied to the sinusoidal waveform shown in Figure 3.21.

$$V_{AVERAGE} = \frac{V_1 + V_2 + V_3 + V_4 + V_5 + V_6}{6}$$

$$= 0.637 \, V_{MAX}$$

For any sinusoidal waveform the average value is equal to 0.637 of the maximum value.

The rms value is the square root of the mean of the individual squared values and is the value of an a.c. voltage which produces the same heating effect as a d.c. voltage. The value can be found from the following formulae applied to the sinusoidal waveform shown in Figure 3.21.

$$V_{rms} = \sqrt{\frac{V_1^2 + V_2^2 + V_3^2 + V_4^2 + V_5^2 + V_6^2}{6}}$$

$$= 0.7071 \, V_{MAX}$$

For any sinusoidal waveform the rms value is equal to 0.7071 of the maximum value.

Example

The sinusoidal waveform applied to a particular circuit has a maximum value of 339.5 V. Calculate the average and rms value of the waveform.

$$\text{Average value } V_{av} = 0.637 \times V_{max}$$
$$\therefore V_{av} = 0.637 \times 339.5 = 216.3 \text{ V}$$
$$\text{rms value} = V_{rms} = 0.7071 \times V_{max}$$
$$V_{rms} = 0.7071 \times 339.5$$
$$= 240 \text{ V}$$

When we say that the main supply to a domestic property is 240 V we really mean 240 V rms. Such a waveform has an average value of about 216 V and a maximum value of almost 340 but because the rms value gives the d.c. equivalent value we almost always give the rms value without identifying it as such.

Resistance and reactance in an a.c. circuit

Resistance (*R*)

In any d.c. circuit the opposition to current flow is called the resistance of the circuit, measured in ohms and given by the symbol R

$$R = \frac{V_R}{I_R} \, (\Omega)$$

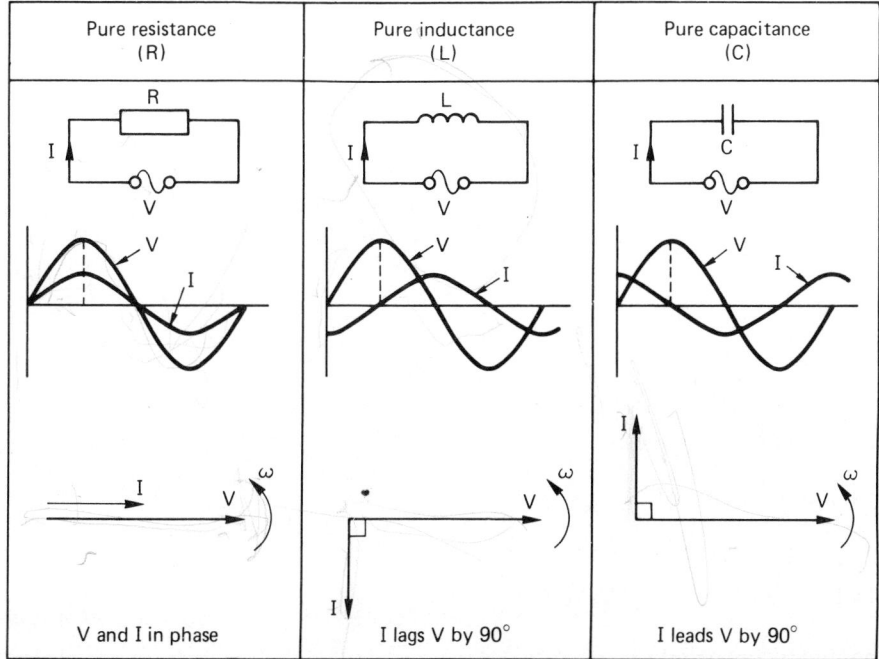

Figure 3.22 Voltage and current relationships in resistive, capacitive and inductive circuits

In an a.c. circuit the total opposition is due to the resistance *and* reactance of the circuit.

Inductive reactance (X_L)

This is the opposition to an a.c. current in an inductive circuit. It causes the current in the circuit to lag behind the applied voltage as shown in Figure 3.22.

Inductive reactance $X_L = \dfrac{V_L}{I_L}\,(\Omega)$

or $X_L = 2\pi f L\,(\Omega)$

where 2 = a number constant
  π = 3.142 a constant
  f = the frequency of the supply
  L = the inductance of the circuit

Capacitive reactance (X_c)

This is the opposition to an a.c. current in a capacitive circuit. It causes the current in the circuit to lead ahead of the voltage as shown in Figure 3.22.

Capacitive reactance $X_C = \dfrac{V_C}{I_C}\,(\Omega)$

or $X_C = \dfrac{1}{2\pi f C}\,(\Omega)$

where 2 = a number constant
  π = 3.142 a constant
  f = the frequency of the supply
  C = the capacitance of the circuit

When circuits contain two or more separate elements such as RL, RC or RLC, the total opposition to current flow is known as the impedance of the circuit and given the symbol Z.

Example 1

Calculate the reactance of a 150 μF capacitor and a 0.05 H inductor if they were separately connected to the 50 Hz mains supply.

For capacitive reactance $X_C = \dfrac{1}{2\pi fC}$

where $f = 50$ Hz and

$C = 150\ \mu\text{F} = 150 \times 10^{-6}\ \text{F}$.

$$\therefore X_C = \frac{1}{2 \times 3.142 \times 50\ \text{Hz} \times 150 \times 10^{-6}}$$

$$= 21.2\ \Omega.$$

For inductive reactance $X_L = 2\pi fL$

where $f = 50$ Hz and $L = 0.05$ H

$\therefore X_L = 2 \times 3.142 \times 50\ \text{Hz} \times 0.05\ \text{H} = 15.7\ \Omega.$

Resistance, inductance and capacitance in an a.c. circuit

When a resistor only is connected to an a.c. circuit the current and voltage waveforms remain together, starting and finishing at the same time. We say that the waveforms are *in phase*.

When a pure inductor is connected to an a.c. circuit the current lags behind the voltage waveform by an angle of 90°. We say that the current *lags* the voltage by 90°. When a pure capacitor is connected to an a.c. circuit the current *leads* the voltage by an angle of 90°. The various effects can be observed on an oscilloscope, but the circuit diagram, waveform diagram and phasor diagram for each circuit is shown in Figure 3.22.

Phasor diagrams

Phasor diagrams and a.c. circuits are an inseparable combination. Phasor diagrams allow us to produce a model or picture of the circuit under consideration, which helps us to understand the circuit. A phasor is a straight line, having definite length and direction, which represents to scale, the magnitude and direction of a quantity such as a current, voltage or impedance.

To find the combined effect of two quantities we combine their phasors by adding the beginning of the second phasor to the end of the first. The combined effect of the two quantities is shown by the resultant phasor, which is measured from the original zero position to the end of the last phasor.

Example 1

Find by phasor addition the combined effect of currents A and B acting in a circuit. Current A has a value of 4 A and current B a value of 3 A, leading A by 90°. We usually assume phasors to rotate anti-clockwise and so the complete diagram will be as shown in Figure 3.23. Choose a scale of, for example, 1 A = 1 cm and draw the phasors to scale, i.e. A = 4 cm and B = 3 cm leading A by 90°.

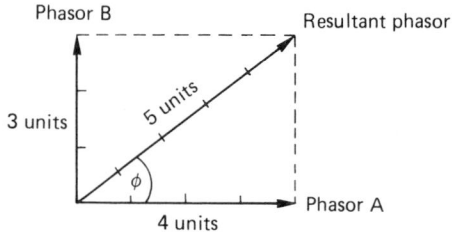

Figure 3.23 The phasor addition of currents A and B

The magnitude of the resultant phasor can be measured from the phasor diagram and is found to be 5 A acting at a phase angle ϕ of about 37° leading A. We, therefore, say that the combined effect of currents A and B is a current of 5 A at an angle of 37° leading A.

Phase angle ϕ

In an a.c. circuit containing resistance only, such as heating circuits, the voltage and current are in phase, that means that they reach their peak and zero values together, as shown in Figure 3.24(a).

In an a.c. circuit containing inductance, such as a motor or discharge lighting circuit, the current often reaches its maximum value after the voltage, this means that the current and voltage are out of phase with each other, as shown in Figure 3.24(b). The phase difference, measured in degrees between the current and voltage, is called the phase angle of the circuit, which is denoted by the symbol ϕ, the small Greek letter phi.

Power factor

The cosine of the phase angle is known as the power factor (p.f.) of the circuit i.e. cos. ϕ = p.f. If the current leads the voltage we say that power

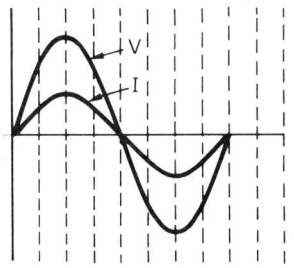

a) V and I in phase
    and, therefore,
    phase angle $\phi = 0°$
    $\cos \phi$ = p.f. = 1

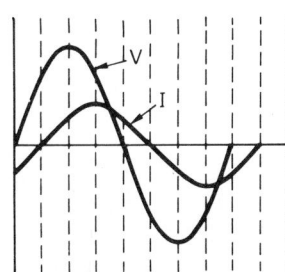

b) V and I displaced
    by 45° and,
    therefore, phase
    angle $\phi = 45°$
    $\cos \phi$ = p.f. = 0.707

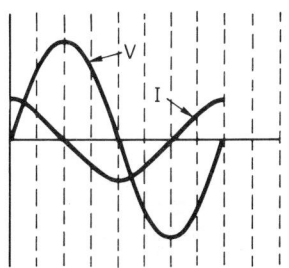

c) V and I displaced
    by 90° and,
    therefore, phase
    angle $\phi = 90°$
    $\cos \phi$ = p.f. = 0

Figure 3.24 Phase relationship of a.c. waveform

factor is leading, as shown in Figure 3.24(c). If the current lags the voltage we say that the power factor is lagging, as shown in Figure 3.24(b). This means that we are using the supply voltage as the reference quantity, which seems sensible since the supply authority maintains a constant voltage, but the current depends upon the load. The ideal situation is when V and I are in phase, reaching their maximum and zero values together. The phase angle is then 0° and the power factor is 1 as shown in Figure 3.24(a).

The power factor of most industrial loads is lagging because the machines and discharge lighting used in industry are mostly inductive. This causes an additional magnetising current to be drawn from the supply, which does not produce power, but does need to be supplied, making supply cables larger.

Example 1

A 250 V supply feeds three 2 kW loads with power factors of 1, 0.8 and 0.4. Calculate the current at each power factor.

Current $I = \dfrac{P}{V \cos \phi}$

where $P = 2\,\text{kW} = 2000\,\text{W}$

and $V = 250\,\text{V}$

at p.f. = 1

$\therefore$ $I = \dfrac{2000\,\text{W}}{250\,\text{V} \times 1} = 8\,\text{A}$

at p.f. = 0.8

$\therefore$ $I = \dfrac{2000\,\text{W}}{250\,\text{V} \times 0.8} = 10\,\text{A}$

at p.f. = 0.4

$\therefore$ $I = \dfrac{2000\,\text{W}}{250\,\text{V} \times 0.4} = 20\,\text{A}$

It can be seen from these calculations that a 2 kW load supplied at a power factor of 0.4 would require a 20 A cable, while the same load at unity power factor could be supplied with an 8 A cable. There may also be the problem of higher voltage drops in the supply cables. As a result, the supply authorities encourage installation engineers to improve their power factor to a value close to 1 and sometimes charge penalties if the power factor falls below 0.8.

Power-factor improvement

Most installations have a low power factor because of the inductive nature of the load. A capacitor has the opposite effect to an inductor, and, therefore, we can connect a capacitor in parallel with a load

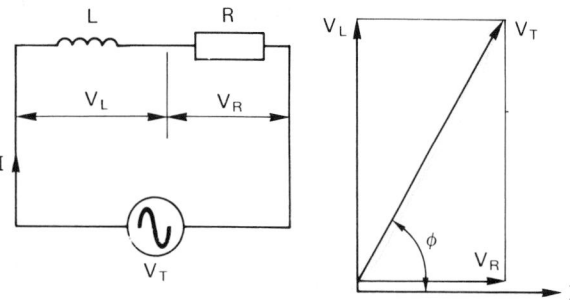

Figure 3.25 A series R-L circuit and phasor diagram

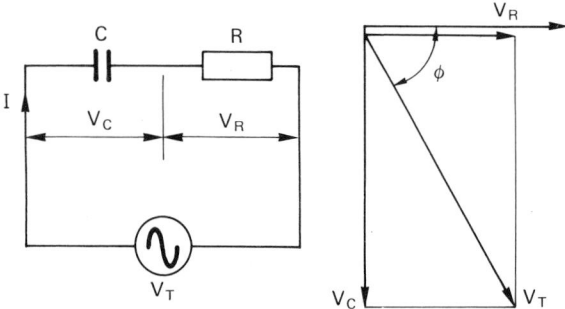

Figure 3.26 A series R-C circuit and phasor diagram

which is known to have a low power factor, in order to improve the power factor of the circuit.

The inductance of the choke in a fluorescent light fitting creates a bad power factor, but by connecting a capacitor across the mains supply to the luminaire, the power factor is improved.

A.c. circuits

In a circuit containing a resistor and inductor connected in series, as shown in Figure 3.25, the current I will flow through the resistor and the inductor, causing the voltages V_R to be dropped across the resistor and V_L to be dropped across the inductor. The sum of these voltages will be equal to the total voltage V_T but as this is an a.c. circuit the voltages must be added by phasor addition. The result is shown in Figure 3.25 where V_R is drawn to scale and in phase with the current and V_L is drawn to scale and leading the current by 90°. The phasor addition of these two voltages gives us the magnitude and direction of V_T which leads the current by some angle ϕ.

In a circuit containing a resistor and capacitor connected in series, as shown in Figure 3.26, the current I will flow through the resistor and capacitor, causing voltage drops V_R and V_C. The voltage V_R will be in phase with the current and V_C will lag the current by 90°. The phasor addition of these voltages is equal to the total voltage V_T which, as can be seen in Figure 3.26, is lagging the current by some angle ϕ.

The impedance triangle

We have now established the general shape of the phasor diagram for a series a.c. circuit. Figures 3.25 and 3.26 show the voltage phasors, but we know that $V_R = IR$, $V_L = IX_L$, $V_C = IX_C$ and $V_T = IZ$, and, therefore, the phasor diagrams (a) and (b) of Figure 3.27 must be equal. From Figure 3.27, by the theorem of Pythagoras, we have

$$(IZ)^2 = (IR)^2 + (IX)^2$$
$$I^2Z^2 = I^2R^2 + I^2X^2$$

If we now divide throughout by I^2 we have

$$Z^2 = R^2 + X^2$$
$$\text{or } Z = \sqrt{R^2 + X^2} \ (\Omega)$$

The phasor diagram can be simplified to the impedance triangle given in Figure 3.27(c).

Example 1

A coil of 0.15 H is connected in series with a 50 Ω resistor across a 100 V 50 Hz supply. Calculate (a) the reactance of the coil, (b) the impedance of the circuit and (c) the current.

For (a) $X_L = 2\pi f L \ (\Omega)$

$\therefore \ X_L = 2 \times 3.142 \times 50 \text{ Hz} \times 0.15 \text{ H}$
$= 47.1 \ \Omega$

For (b) $Z = \sqrt{R^2 + X^2} \ (\Omega)$

$\therefore \ Z = \sqrt{(50 \ \Omega)^2 + (47.1 \ \Omega)^2} = 68.69 \ \Omega$

For (c) $I = V/Z \ (A)$

$\therefore \ I = \dfrac{100 \text{ V}}{68.69 \ \Omega} = 1.46 \text{ A}$

For an inductive circuit

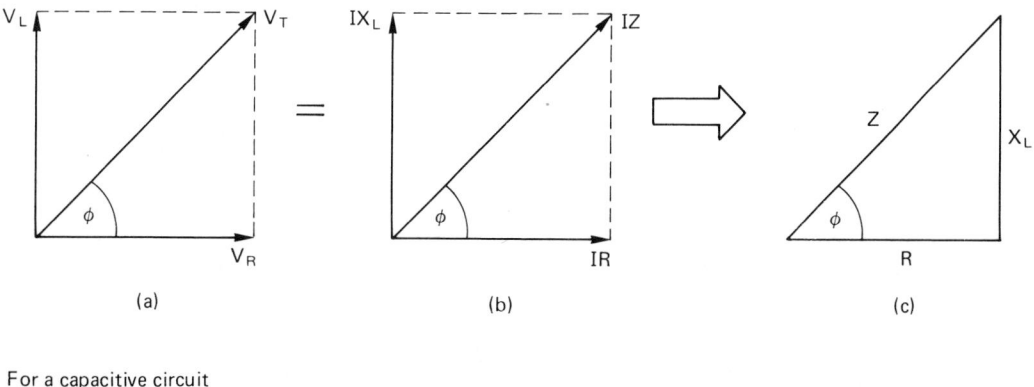

(a)   (b)   (c)

For a capacitive circuit

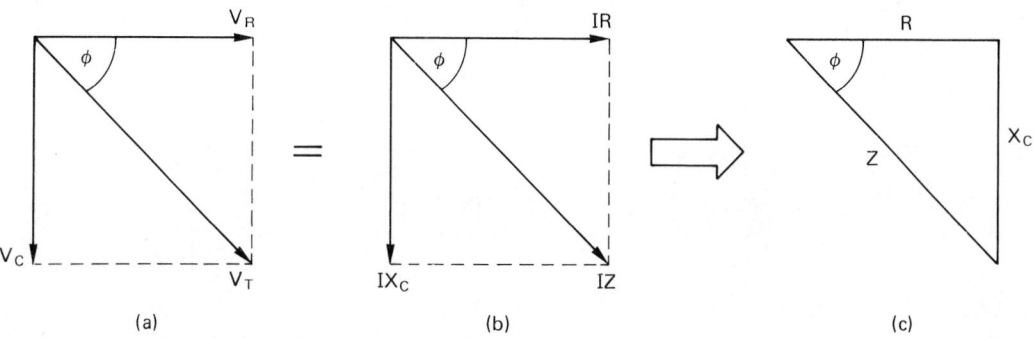

(a)   (b)   (c)

Figure 3.27 Phasor diagram and impedance triangle

Example 2

A 60 μF capacitor is connected in series with a 100 Ω resistor across a 240 V 50 Hz supply. Calculate (a) the reactance of the capacitor, (b) the impedance of the circuit and (c) the current.

For (a) $X_C = \dfrac{1}{2\pi f C}\,(\Omega)$

$\therefore\ X_C = \dfrac{1}{2\pi \times 50\ \text{Hz} \times 60 \times 10^{-6}\ \text{F}}$

$= 53.05\ \Omega$

For (b) $Z = \sqrt{R^2 + X^2}\,(\Omega)$

$\therefore\ Z = \sqrt{(100\ \Omega)^2 + (53.05\ \Omega)^2} = 113.2\ \Omega$

For (c) $I = V/Z\,(\text{A})$

$\therefore\ I = \dfrac{240\ \text{V}}{113.2\ \Omega} = 2.12\ \text{A}$

Power and power factor

A little earlier in this chapter power factor was defined as the cosine of the phase angle between the current and voltage. Power factor may be abbreviated to p.f. If the current lags the voltage, as shown in Figure 3.25, we say that the p.f. is lagging, and if the current leads the voltage as shown in Figure 3.26, the p.f. is said to be leading. p.f = $\cos \phi$.

From the trigonometry of the impedance triangle shown in Figure 3.27, p.f. is also equal to:

$$\text{p.f.} = \cos \phi = \frac{R}{Z} = \frac{V_R}{V_T}$$

The electrical power in a circuit is the product of the instantaneous value of the voltage and current. Figure 3.28 shows the voltage and current waveform for a pure indicator and pure capacitor.

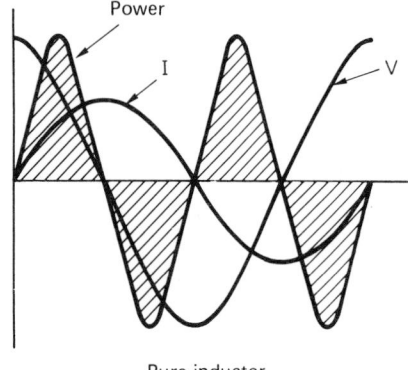

Pure inductor

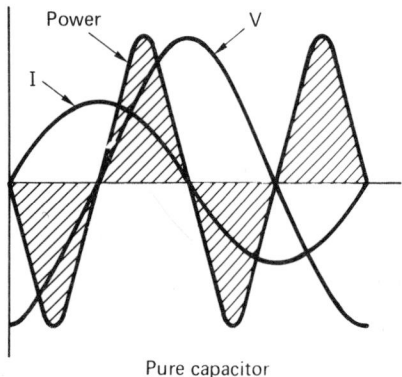

Pure capacitor

Figure 3.28 Waveform for the a.c. power in purely inductive and purely capacitive circuits

The power waveform is obtained from the product of V and I at every instant in the cycle. It can be seen that the power waveform reverses every quarter cycle, indicating that energy is alternately being fed into and taken out of the inductor and capacitor. When considered over one complete cycle, the positive and negative portions are equal, showing that the average power consumed by a pure inductor or capacitor is zero. This shows that inductors and capacitors store energy during one part of the voltage cycle and feed it back into the supply later in the cycle. Inductors store energy as a magnetic field and capacitors as an electric field.

In an electric circuit more power is taken from the supply than is fed back into it, since some power is dissipated by the resistance of the circuit and, therefore,

$$P = I^2 R \text{ (W)}$$

In any d.c. circuit the power consumed is given by the product of the voltage and current, because in a d.c. circuit, voltage and current are in phase. In an a.c. circuit the power consumed is given by the product of the current and that part of the voltage which is in phase with the current. The in-phase component of the voltage is given by V cos ϕ and so power can also be given by the equation

$$P = VI \cos \phi \text{ (W)}$$

Example 1

A coil has a resistance of 30 Ω and a reactance of 40 Ω when connected to a 250 V supply. Calculate (a) the impedance, (b) the current, (c) p.f. and (d) the power.

For (a) $Z = \sqrt{R^2 + X^2} \text{ (}\Omega\text{)}$

$\therefore$ $Z = \sqrt{(30\,\Omega)^2 + (40\,\Omega)^2} = 50\,\Omega$

For (b) $I = V/Z \text{ (A)}$

$\therefore$ $I = \dfrac{250 \text{ V}}{50\,\Omega} = 5 \text{ A}$

For (c)   p.f. $= \cos\phi = \dfrac{R}{Z}$

$\therefore$ p.f. $= \dfrac{30\,\Omega}{50\,\Omega} = 0.6$ lagging

For (d) $P = VI \cos\phi \text{ (W)}$

$\therefore$ $P = 250 \text{ V} \times 5 \text{ A} \times 0.6 = 750 \text{ W}$

Example 2

A capacitor of reactance 12 Ω is connected in series with a 9 Ω resistor across a 150 V supply.

Calculate (a) the impedance of the circuit, (b) the current, (c) the p.f. and (d) the power.

For (a) $\quad Z = \sqrt{R^2 + X^2}\ (\Omega)$

$\therefore \quad\quad Z = \sqrt{(9\ \Omega)^2 + (12\ \Omega)^2} = 15\ \Omega$

For (b) $\quad I = V/Z\ (\mathrm{A})$

$\therefore \quad\quad I = \dfrac{150\ \mathrm{V}}{15\ \Omega} = 10\ \mathrm{A}$

For (c) $\quad$ p.f. $= \cos\phi = \dfrac{R}{Z}$

$\therefore \quad\quad$ p.f. $= \dfrac{9\ \Omega}{15\ \Omega} = 0.6$ leading

For (d) $\quad P = VI\cos\phi\ (\mathrm{W})$

$\therefore \quad\quad P = 150\ \mathrm{V} \times 10\ \mathrm{A} \times 0.6 = 900\ \mathrm{W}$

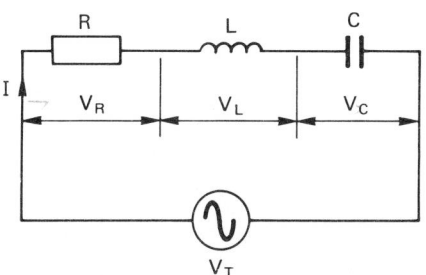

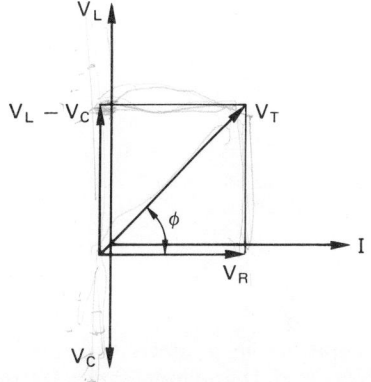

Figure 3.29 R-L-C series circuit and phasor diagram

Resistance, inductance and capacitance in series

The circuit diagram and phasor diagram of an R, L, C series circuit are shown in Figure 3.29. The voltages across the components are represented by V_R, V_L and V_C, which have the directions shown. Since V_L leads I by 90° and V_C lags by 90° the phasors are in opposition and the combined result is given by $V_L - V_C$ as shown.

Applying the theorem of Pythagoras to the phasor diagram of Figure 3.29 we have

$$V_T^2 = V_R^2 + (V_L - V_C)^2$$

Since $V_T = IZ$, $V_R = IR$, $V_L = IX_L$ and $V_C = IX_C$ this equation may also be expressed thus:

$$(IZ)^2 = (IR)^2 + (IX_L - IX_C)^2.$$

Cancelling out the common factors we have

$$Z^2 = R^2 + (X_L - X_C)^2$$
$$\text{or } Z = \sqrt{R^2 + (X_L - X_C)^2}\ (\Omega)$$

Note The derivation of this equation is not required by craft students but the equation should be remembered and applied in appropriate cases.

Example 1

A coil of resistance 5 Ω and inductance 10 mH is connected in series with a 75 μF capacitor across a 200 V 100 Hz supply. Calculate (a) the impedance of the circuit, (b) the current and (c) the p.f.

For (a) $\quad X_L = 2\pi fL\ (\Omega)$
$\therefore \quad\quad X_L = 2 \times 3.142 \times 100\ \mathrm{Hz} \times 10 \times 10^{-3}\ \mathrm{H}$
$\quad\quad\quad X_L = 6.28\ \Omega$

$\quad\quad\quad X_C = \dfrac{1}{2\pi fC}\ (\Omega)$

$\therefore \quad\quad X_C = \dfrac{1}{2 \times 3.142 \times 100\ \mathrm{Hz} \times 75 \times 10^{-6}\ \mathrm{F}}$

$\quad\quad\quad X_C = 21.22\ \Omega$

$\quad\quad\quad Z = \sqrt{R^2 + (X_L - X_C)^2}\ (\Omega)$

$$\therefore \qquad Z = \sqrt{(5\ \Omega)^2 + (6.28\ \Omega - 21.22\ \Omega)^2}$$

$$Z = 15.75\ \Omega$$

For (b) $\quad I = V/Z\ (\text{A})$

$$\therefore \qquad I = \frac{200\ \text{V}}{15.75\ \Omega} = 12.69\ \text{A}$$

For (c) p.f. $= \cos\phi = \dfrac{R}{Z}$

$$\therefore \qquad \text{p.f.} = \frac{5}{12.69} = 0.39$$

Example 2

A 200 μF capacitor is connected in series with a coil of resistance 10 Ω and inductance 100 mH to a 240 V 50 Hz supply. Calculate (a) the impedance, (b) the current and (c) the voltage dropped across each component.

For (a) $X_L = 2\pi f L\ (\Omega)$

$$\therefore \qquad X_L = 2 \times 3.142 \times 50\ \text{Hz} \times 100 \times 10^{-3}\ \text{H}$$

$$X_L = 31.42\ \Omega$$

$$X_C = \frac{1}{2\pi f C}\ (\Omega)$$

$$\therefore \qquad X_C = \frac{1}{2 \times 3.142 \times 50\ \text{Hz} \times 200 \times 10^{-6}}$$

$$X_C = 15.9\ \Omega$$

$$Z = \sqrt{R^2 + (X_L - X_C)^2}\ (\Omega)$$

$$\therefore \qquad Z = \sqrt{(10\ \Omega)^2 + (31.42\ \Omega - 15.9\ \Omega)^2}$$

$$Z = 18.46\ \Omega$$

For (b) $\quad I = V/Z\ (\text{A})$

$$\therefore \qquad I = \frac{240\ \text{V}}{18.46\ \text{A}} = 13\ \text{A}$$

For (c) $\quad V_R = I \times R\ (\text{V})$
$$\therefore \qquad V_R = 13\ \text{A} \times 10\ \Omega = 130\ \text{V}$$
$$V_L = I \times X_L\ (\text{V})$$
$$\therefore \qquad V_L = 13\ \text{A} \times 31.42\ \Omega = 408.46\ \text{V}$$
$$V_C = I \times X_C\ (\text{V})$$
$$\therefore \qquad V_C = 13\ \text{A} \times 15.9\ \Omega = 206.7\ \text{V.}$$

The phasor diagram of this circuit would be similar to that shown in Figure 3.29.

Series resonance

At resonance the circuit responds sympathetically. Therefore, the condition of resonance is used extensively in electronic and communication circuits for frequency selection and tuning. The current and reactive components of the circuit are at a maximum and so resonance is usually avoided in power applications to prevent cables being overloaded and cable insulation being broken down.

A circuit can be tuned to resonance by either varying the capacitance of the circuit or by adjusting the frequency. At low frequencies the circuit is mainly capacitive and at high frequencies the inductive effect predominates. At some intermediate frequency a point exists where the capacitive effect exactly cancels the inductive effect. This is the point of resonance and occurs when

$$V_L = V_C$$
$$\therefore \qquad I X_L = I X_C$$

If we cancel the common factor we have

$$X_L = X_C$$

$$2\pi f L = \frac{1}{2\pi f C}$$

Collecting terms

$$f^2 = \frac{1}{4\pi^2 LC}$$

Taking square roots,

Resonant frequency $= f_0 = \dfrac{1}{2\pi}\sqrt{\dfrac{1}{LC}}$ (Hz)

Note   The resonant frequency is given the symbol f_0. The derivation of the formulae is not required by craft students.

At resonance the circuit is purely resistive. $Z = R$, the phase angle is zero and, therefore, the supply voltage and current must be in phase. These effects are shown in Figure 3.30.

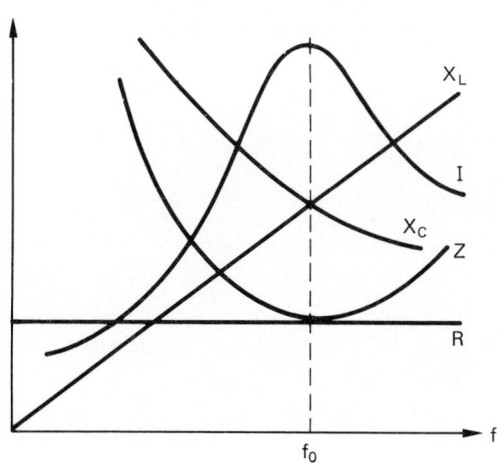

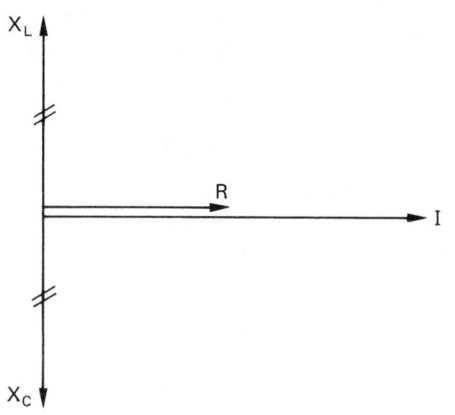

Figure 3.30 Series resonance conditions in an R-L-C circuit

Example 1

A capacitor is connected in series with a coil of resistance 50 Ω and inductance 168.8 mH across a 50 Hz supply. Calculate the value of the capacitor to produce resonance in this circuit.

$X_L = 2\pi fL$ (Ω)
$X_L = 2 \times 3.142 \times 50 \text{ Hz} \times 168.8 \times 10^{-3}$ H.
$X_L = 53.03$ Ω.

At resonance $X_L = X_C$ therefore $X_C = 53.03$ Ω

$$X_C = \frac{1}{2\pi fC} \ (\Omega)$$

Transposing for $C = \dfrac{1}{2\pi fX_C}$ (F)

$$\therefore C = \frac{1}{2 \times 3.142 \times 50 \text{ Hz} \times 53.03 \ \Omega}$$

$$C = 60 \ \mu\text{F}$$

Example 2

Calculate the resonant frequency of a circuit consisting of a 25.33 mH inductor connected in series with a 100 μF capacitor.

Resonant frequency $= f_0 = \dfrac{1}{2\pi}\sqrt{\dfrac{1}{LC}}$ (Hz)

$$\therefore f_0 = \frac{1}{2\pi}\sqrt{\frac{1}{25.33 \times 10^{-3} \text{ H} \times 100 \times 10^{-6} \text{ F}}}$$

$$f_0 = 100 \text{ Hz}.$$

Transformers

A transformer is an electrical machine which is used to change the value of an alternating voltage. They vary in size from miniature units used in electronics to huge power transformers used in power stations. A transformer will only work when an alternating voltage is connected. It will not normally work from a d.c. supply such as a battery.

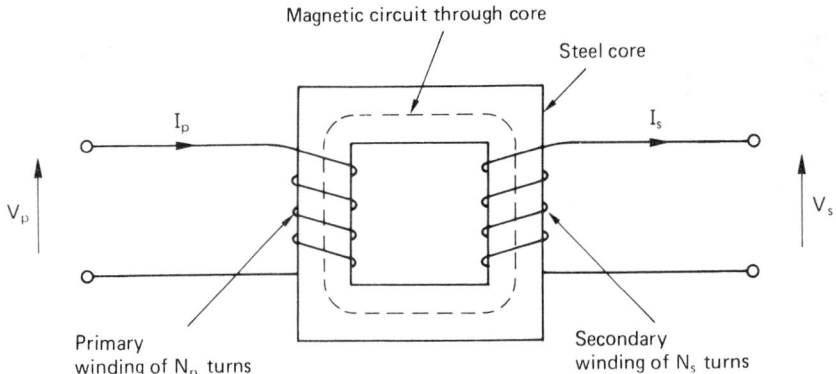

Figure 3.31 A simple transformer

A transformer, as shown by the diagram in Figure 3.31, contains two coils, called the primary and secondary coils, or windings, which are insulated from each other and wound on the same steel or iron core.

An alternating voltage applied to the primary winding produces an alternating current, which sets up an alternating magnetic flux throughout the core. This magnetic flux induces an e.m.f. into the secondary winding as described by Faraday's Laws, which say '*when a conductor is cut by a magnetic field, an e.m.f. is induced in that conductor*'. Since both windings are linked by the same magnetic flux, the induced e.m.f. per turn will be the same for both windings. Therefore, the e.m.f. in both windings is proportional to the number of turns.

Writing this expression as an equation we have:

The volts per turn on the primary winding = The volts per turn on the secondary winding

If we now write this equation as a formula we have:

$$\frac{V_P}{N_P} = \frac{V_S}{N_S} \qquad \text{(Equation 1)}$$

Most practical power transformers have a very high efficiency and for an ideal transformer having a one hundred per cent efficiency the primary power is equal to the secondary power.

Primary power = Secondary power
and since
Power   = Voltage × Current

then $V_P \times I_P = V_S \times I_S$ (Equation 2)

Combining Equations 1 and 2 we have:

$$\frac{V_P}{V_S} = \frac{N_P}{N_S} = \frac{I_S}{I_P}$$

Example 1

A 240 V to 12 V bell transformer is constructed with 800 turns on the primary winding. Calculate the number of secondary turns and the primary and secondary currents when the transformer supplies a 12 V, 12 W alarm bell.

Collecting the information given in the question into a usable form we have:

$$V_P = 240 \text{ V}$$
$$V_S = 12 \text{ V}$$
$$N_P = 800 \text{ turns}$$
Power = 12 W

Information required: N_S, I_S and I_P
Secondary turns

$$N_S = \frac{N_P V_S}{V_P}$$

$$\therefore N_S = \frac{800 \text{ turns} \times 12 \text{ V}}{240 \text{ V}} = 40 \text{ turns.}$$

Secondary current

$$I_S = \frac{\text{POWER}}{V_S}$$

$$\therefore \quad I_S = \frac{12\text{ W}}{12\text{ V}} = 1\text{ A}$$

Primary current

$$I_P = \frac{I_S \times V_S}{V_P}$$

$$\therefore \quad I = \frac{1\text{ A} \times 12\text{ V}}{240\text{ V}} = 0.05\text{ A}$$

Transformer losses

Transformers have a very high efficiency, usually better than 90% because they have no moving parts causing frictional losses. However, the losses which do occur in a transformer can be grouped under two general headings; copper losses and iron losses.

Copper losses occur because of the small internal resistance of the windings. They are proportional to the load, increasing as the load increases because copper loss is an 'I^2R' loss.

Iron losses are made up of *hysteresis loss* and *eddy current loss*. The hysteresis loss depends upon the type of iron used to construct the core and consequently core materials are carefully chosen. Transformers will only operate on an alternating current, thus the current which establishes the core flux is constantly changing from positive to negative. Each time there is a current reversal, the magnetic flux reverses and it is this building up and collapse of magnetic flux in the core material which accounts for the hysteresis loss.

Eddy currents are circulating currents created in the core material by the changing magnetic flux.

The iron loss is a constant loss consuming the same power from no load to full load.

Transformer construction

Transformers are constructed in a way which reduces the losses to a minimum. The core of a power transformer is usually made of silicon-iron laminations, because at fixed low frequencies

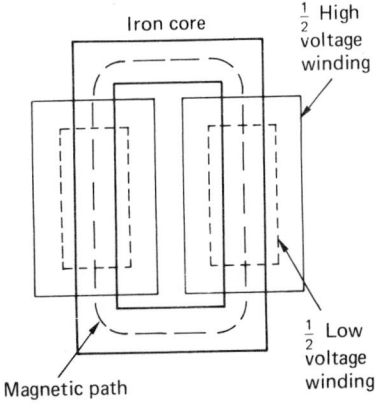

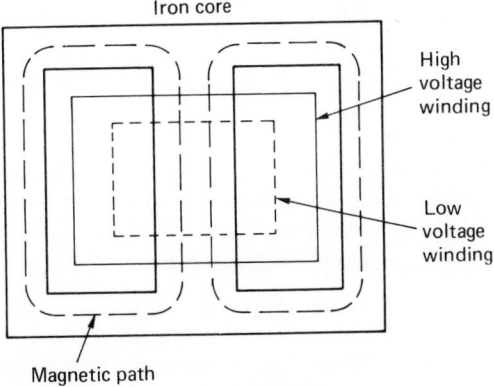

Figure 3.32 Transformer construction

silicon-iron has a small hysteresis loss and the laminations reduce the eddy current loss.

At radio frequencies and above, steel cores cannot be used because the losses become excessive. One solution has been to form *Ferrite* cores for transformers used in electronic circuits. Ferrite cores are made by suspending iron dust in a non-conducting bakelite medium.

The primary and secondary windings are wound close to each other on the central limb of the transformer core. If the windings are spread over two limbs, there will usually be half of each winding on each limb as shown in Figure 3.32.

Matching transformers

Transformers are sometimes used as an impedance matching device between a load of low impedance, such as an audio speaker, and an amplifier of high

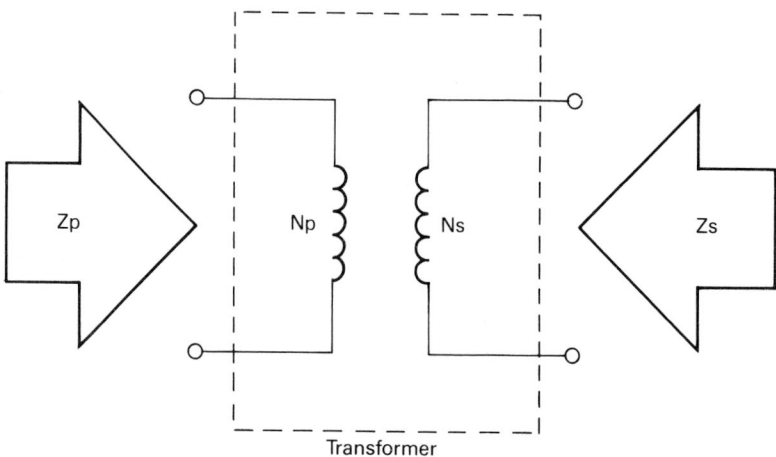

Figure 3.33 Matching transformer connections

output impedance. The purpose of the transformer is to make the load presented to the output terminals of the amplifier '*look*' as though it has the same impedance as the amplifier because this ensures maximum power transfer between the amplifier and the load.

This arrangement is shown in Figure 3.33, where

Z_P is the amplifier output presented to the primary terminals of the transformer

Z_S is the load impedance connected to the secondary terminals of the transformer

N_P is the number of primary turns

N_S is the number of secondary turns

For a matching transformer

$$\frac{N_P}{N_S} = \sqrt{\frac{Z_P}{Z_S}}$$

Example 1

The output impedance of an audio amplifier is 800 Ω and it is proposed to connect an 8 Ω speaker to this amplifier. Determine the turns ratio of the transformer which would ensure maximum power transfer between the amplifier and the speaker.

$$\frac{N_P}{N_S} = \sqrt{\frac{Z_P}{Z_S}}$$

$$\therefore \frac{N_P}{N_S} = \sqrt{\frac{800\,\Omega}{8\,\Omega}} = 10$$

Therefore, a transformer having a turns ratio of 10:1 is required.

Example 2

A transformer with a turns ratio of 20:1 is to be used to connect a speaker to a 2 kΩ amplifier. Determine the value of the speaker which will produce maximum power transfer.

$$\frac{N_P}{N_S} = \sqrt{\frac{Z_P}{Z_S}}$$

$$\therefore \frac{20}{1} = \sqrt{\frac{2\,k\Omega}{Z_S}}$$

$$\frac{20^2}{1} = \frac{2\,k\Omega}{Z_S}$$

$$Z_S = \frac{2\,k\Omega}{20^2}$$

$$Z_S = \frac{2000\,\Omega}{400}$$

$$Z_S = 5\,\Omega$$

A 5 Ω speaker will ensure maximum power is transferred from the amplifier.

Exercises

1. The SI units of length, resistance and power are:
(a) millimetre, ohm, kilowatt (b) centimetre, ohm, watt (c) metre, ohm, watt (d) kilometre, ohm, kilowatt.

2. The current taken by a $10 \, \Omega$ resistor when connected to a 240 V supply is:
(a) 41 mA (b) 2.4 A (c) 24 A (d) 240 A.

3. The resistance of an element which takes 12 A from a 240 V supply is:
(a) $2.88 \, \Omega$ (b) $5 \, \Omega$ (c) $12.24 \, \Omega$ (d) $20 \, \Omega$.

4. A $12 \, \Omega$ lamp was found to be taking a current of 2 A at full brilliance. The voltage across the lamp under these conditions was:
(a) 6 V (b) 12 V (c) 24 V (d) 240 V.

5. The resistance of 100 m of 1 mm^2 cross-section copper cable of resistivity $17.5 \times 10^{-9} \, \Omega$m will be:
(a) $1.75 \, \text{m}\Omega$ (b) $1.75 \, \Omega$ (c) $17.5 \, \Omega$ (d) $17.5 \, \text{k}\Omega$.

6. The resistance of a motor field winding at 0°C was found to be $120 \, \Omega$. Find its new resistance at 20°C if the temperature co-efficient of the winding is $0.004 \, \Omega/\Omega$°C.
(a) $116.08 \, \Omega$ (b) $120.004 \, \Omega$ (c) $121.08 \, \Omega$ (d) $140.004 \, \Omega$.

7. The resistance of a motor field winding was found to be $120 \, \Omega$ at an ambient temperature of 20°C. If the temperature co-efficient of resistance is $0.004 \, \Omega/\Omega$°C the resistance of the winding at 60°C will be approximately:
(a) $102 \, \Omega$ (b) $120 \, \Omega$ (c) $130 \, \Omega$ (d) $138 \, \Omega$.

8. A capacitor is charged by a steady current of 5 mA for 10 s. The total charge stored on the capacitor will be:
(a) 5 mC (b) 50 mC (c) 5 C (d) 50 C

9. When 100 V was connected to a $20 \, \mu$F capacitor the charge stored was:
(a) 2 mC (b) 5 mC (c) 20 mC (d) 100 mC.

10. Resistors of $6 \, \Omega$ and $3 \, \Omega$ are connected in series. The combined resistance value will be:
(a) $2 \, \Omega$ (b) $3.6 \, \Omega$ (c) $6.3 \, \Omega$ (d) $9 \, \Omega$.

11. Resistors of 3 ohm and 6 ohm are connected in parallel. The equivalent resistance will be:
(a) $2 \, \Omega$ (b) $3.6 \, \Omega$ (c) $6.3 \, \Omega$ (d) $9 \, \Omega$.

12. Three resistors of $24 \, \Omega$, $40 \, \Omega$ and $60 \, \Omega$ are connected in series. The total resistance will be:
(a) $12 \, \Omega$ (b) $26.4 \, \Omega$ (c) $44 \, \Omega$ (d) $124 \, \Omega$.

13. Resistors of $24 \, \Omega$, $40 \, \Omega$ and $60 \, \Omega$ are connected together in parallel. The effective resistance of this combination will be:
(a) $12 \, \Omega$ (b) $26.4 \, \Omega$ (c) $44 \, \Omega$ (d) $124 \, \Omega$.

14. Two identical resistors are connected in series across a 12 V battery. The voltage drop across each resistor will be:
(a) 2 V (b) 3 V (c) 6 V (d) 12 V.

15. Two identical resistors are connected in parallel across a 24 V battery. The volt drop across each resistor will be:
(a) 6 V (b) 12 V (c) 24 V (d) 48 V.

16. A $6 \, \Omega$ resistor is connected in series with a $12 \, \Omega$ resistor across a 36 V supply. The current flowing through the $6 \, \Omega$ resistor will be:
(a) 2 A (b) 3 A (c) 6 A (d) 9 A.

17. A $6 \, \Omega$ resistor is connected in parallel with a $12 \, \Omega$ resistor across a 36 V supply. The current flowing through the $12 \, \Omega$ resistor will be:
(a) 2 A (b) 3 A (c) 6 A (d) 9 A.

18. Capcaitors of $24 \, \mu$F, $40 \, \mu$F and $60 \, \mu$F are connected in series. The equivalent capacitance will be:
(a) $12 \, \mu$F (b) $44 \, \mu$F (c) $76 \, \mu$F (d) $124 \, \mu$F.

19. Capacitors of $24 \, \mu$F $40 \, \mu$F and $60 \, \mu$F are connected in parallel. The total capacitance will be:
(a) $12 \, \mu$F (b) $44 \, \mu$F (c) $76 \, \mu$F (d) $124 \, \mu$F.

20. The total power dissipated by a $6 \, \Omega$ and $12 \, \Omega$ resistor connected in parallel across a 36 V supply will be:
(a) 72 W (b) 324 W (c) 576 W (d) 648 W.

21. Three resistors are connected in series and a

current of 10 A flows when they are connected to a 100 V supply. If another resistor of 10 Ω is connected in series with the three series resistors, the current carried by this resistor will be:
(a) 4 A (b) 5 A (c) 10 A (d) 100 A.

22. The rms value of a sinusoidal waveform whose maximum value is 100 V will be:
(a) 63.7 V (b) 70.71 V (c) 100 V (d) 100.67 V.

23. The average value of a sinusoidal alternating current whose maximum value is 10 A will be:
(a) 6.37 A (b) 7.071 A (c) 10 A (d) 10.67 A.

24. The capacitive reactance of a 100 μF capacitor connected to the mains supply will be:
(a) 0.314 Ω (b) 31.83 Ω (c) 5000 Ω (d) 31 kΩ.

25. The inductive reactance of a 0.10 H inductor connected to the mains supply will be:
(a) 0.314 Ω (b) 31.42 Ω (c) 31.83 Ω (d) 3142 Ω.

26. Two a.c. voltages V_1 and V_2 have values of 20 and 30 volts respectively. If V_1 leads V_2 by 45° the resultant voltage will be:
(a) 16 V at 24° (b) 45 V at 90° (c) 46 V at 18° (d) 50 V at 45°.

27. The transformation ratio of a step-down transformer is 20:1. If the primary voltage is 240 V the secondary voltage will be:
(a) 2.4 V (b) 12 V (c) 20 V (d) 24 V.

28. A sinusoidal alternating voltage has a maximum value of 340 V. Draw to scale one full cycle of this waveform and find (a) the instantaneous value of the voltage after 60°, (b) the rms value of the voltage using the mid-ordinate rule and (c) the average value of the voltage using the mid-ordinate rule.

29. Describe the structure of a material which is classified as:
(a) a good conductor and (b) a good insulator.

30. Describe with sketches the meaning of the terms *frequency* and *period* as applied to an a.c. waveform.

31. Describe with the aid of phasor diagram sketches, the meaning of (a) a bad power factor, (b) a good power factor and (c) explain how a bad power factor may be improved.

32. Describe the construction of a small transformer designed to reduce losses when operating (a) at mains frequencies only and (b) at radio frequencies.

33. Sketch the voltage and current phasor diagrams for a series circuit containing:
(a)   resistance and inductance,
(b)   resistance and capacitance
(c)   resistance, inductance and capacitance.

34. Sketch the phasor and waveform diagrams for a series circuit at resonance.

35. State one advantage and one disadvantage of a series resonant circuit.

36. Calculate the resonant frequency when an inductor of 1 mH is connected in series with a 1 μF capacitor.

37. Use a neat sketch to show how an ammeter and voltmeter are connected to a simple series circuit to measure total current and total voltage.

38. Calculate the turns ratio of a matching transformer to ensure maximum power transfer from a 5k Ω amplifier to an 8 Ω speaker.

Solutions

1. c.	11. a.	21. b.	29. answers in text.
2. c.	12. d.	22. b.	30. answers in text.
3. d.	13. a.	23. a.	31. answers in text.
4. c.	14. c.	24. b.	32. answers in text.
5. b.	15. c.	25. b.	33. answers in text.
6. c.	16. a.	26. c.	34. answers in text.
7. d.	17. b.	27. b.	35. answers in text.
8. b.	18. a.	28. (a) =295 V.	36. 5.03 kHz.
9. a.	19. d.	(b) =240 V.	37. answers in text.
10. d.	20. b.	(c) =217 V.	38. 25:1.

CHAPTER 4

Electronic systems

The word *system* has only recently become popular and assumed importance as a result of the widespread use of the term. Our world is a part of the *solar system*, we live in a *social system*, doctors study the *nervous system* and scientists study the *ecological system*. In general, the theory of systems looks at the things which are common or related in an effort to make the overall behaviour more understandable. This moves us towards a definition of the system. 'A system is a collection of parts which are joined or connected together in some particular way.'

A systems approach is a procedure or strategy which is used as a way of finding solutions to a complex problem. This involves using a system diagram or block diagram linked by interconnecting lines and arrows.

A *system* or *block diagram* is made up of a number of boxes, each box representing a part of the system which can also be called a sub-system. The purpose of the block diagram is to show how the various parts or sub-systems relate to, or interconnect with, each other. These boxes are often called *black boxes* because initially we don't want to see what is inside the box. We only want to understand the overall behaviour and the relationship of one box with another.

The interconnections between the various parts of the system, or between the black boxes, is represented by lines and arrows, the arrows indicating the direction of flow.

This approach to problem solving allows us to concentrate our efforts on understanding the system without getting confused by the circuit complexities of each black box.

A cassette tape recorder can be used to store sound on a magnetic tape and, some time later, to play back those sounds through a speaker. A tape recorder is a sophisticated piece of electronic equipment made up of complex circuits, but if we look at it as a system which is made up of black boxes, we can begin to understand how it works without necessarily being able to understand the individual parts of the circuit. Figure 4.1 shows a system diagram for a cassette tape recorder.

The first black box represents the microphone which picks up the sound waves and converts them into an electrical signal. The amplifier, box two, modifies those signals and stores them on the magnetic tape, box three. On playback the magnetic tape signal must be amplified, box four, before being played back through the speakers. Therefore, the tape recorder has five sub-systems or black boxes. Each box has a different function and in reality will be made up of complicated electromechanical components and circuits. Even though we don't yet know how each block will be constructed, we have made a start in our understanding of how the tape recorder works. A great deal can be done in electronics if you know what a particular black box should do, even though you don't know how it does it.

In electronics, engineering, instrumentation and control, we are usually using an electronic system to control something for a purpose. For example, to control water temperature or motor speed or fluid levels or the output of an audio system. Whatever their level of sophistication, all control systems have certain common basic features. The simplest form of control is open loop control.

Open loop control

Open loop control involves designing a system to do a particular job as carefully as possible and then leaving it to work on its own. With open loop control you are not in a position to make correc-

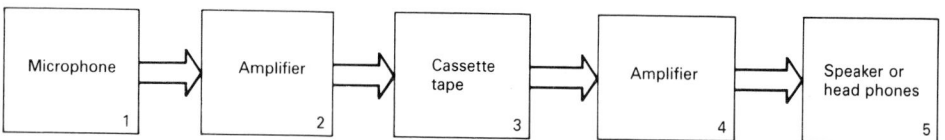

Figure 4.1 A system diagram for a tape recorder

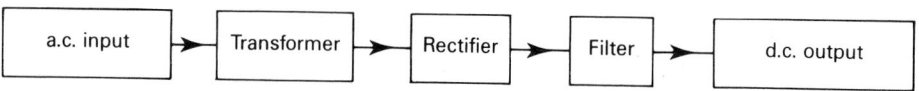

Figure 4.2 D.c. power supply

tions once the system is set up and working. Consider the following example. A golfer stands suitably poised, holding the club and ready to take a shot. He addresses the ball and, after swinging back the club with great care, strikes the ball. When the ball has left the club face the golfer has lost all control over the flight of the ball and further corrections cannot be made. If the shot is not going where it was planned it should go, then all that can be done is to take another ball and try again. This is, therefore, an example of open loop control.

Consider the d.c. power supply shown in Figure 4.2. The transformer reduces the a.c. mains voltage to a lower a.c. voltage. The rectifier converts this a.c. voltage into a d.c. voltage by connecting four diodes in a bridge circuit. The output from a bridge circuit is lumpy d.c. and therefore the filter is required to smooth the output. This will probably be an electrolytic capacitor connected across the output. The d.c. power supply will have been carefully designed and the components carefully selected to give a specified output, but if any of the variables change, there is no facility to make corrections and, therefore, this too is an example of open loop control.

Closed loop control

For closed loop control there must be a way of feeding back information so that adjustments can be made to correct errors in the system. Driving a motor car is a good example of closed loop control. The driver constantly makes corrections to speed

and direction in response to observations of the changing road conditions.

Motor speed control

The motor speed controller shown in Figure 4.3 is also an example of closed loop control. The error detector has two inputs and one output. The desired speed is set by the speed control which supplies a reference signal to the error detector. The output from the error detector supplies a signal to the power amplifier, which provides the necessary power to drive the motor. The drive shaft of the motor, in addition to driving the mechanical load, also turns a tacho generator. This is a small generator coupled to the drive shaft, which generates a voltage which is proportional to the shaft speed. Increasing the shaft speed increases the generated voltage, reducing the speed reduces the generated voltage. This voltage is then fed back to the error detector, which compares the feedback signal with the reference signal. The difference between the reference signal and the feedback signal, the error, is then used to adjust the output to the power amplifier, so that the chosen speed can be maintained.

This system is, therefore, made up of five black boxes or sub-systems. The set speed control, which provides the reference signal, might be a voltage set by the wiper of a variable resistor or potentiometer. The error detector could be an operational amplifier because op-amps can compare the voltage at their two inputs. The power amplifier will probably be an electronic circuit built up from

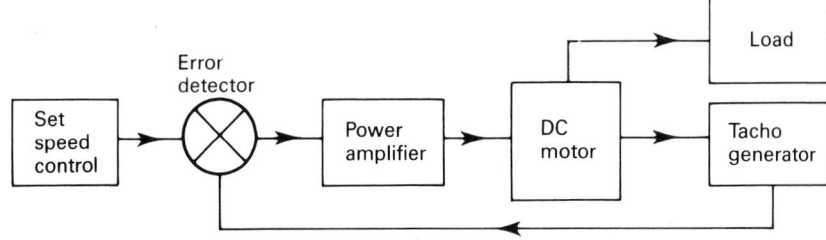

Figure 4.3 Motor speed control

discrete components as shown in Figure 4.4 and considered in more detail in Chapter 7. Low power systems can be made into integrated circuits and cooled by mounting on heat sinks, but high power systems tend to be made with discrete components because they generate a lot of heat and need to be bulky to provide the necessary cooling of the components. The power amplifier drives the motor which, in turn, drives the tacho generator. This generates the feedback voltage, which is connected to the input of the error detector, forming the closed loop in this system.

The feedback signal of a closed loop system can be either positive or negative.

Negative feedback

Negative feedback is the name given to the situation in which the feedback signal is subtracted from the reference signal with the intention of

Figure 4.4 Using heat sink and discrete components to cool a power amplifier

reducing the error or variations in the system. Negative feedback is the type normally used in control systems because it leads to stability. The motor speed control system described above and shown in Figure 4.3 will incorporate negative feedback because the system is designed to reduce variations in the speed of the motor. Electronic amplifier circuits normally incorporate negative feedback because the voltage gain of the amplifier can be accurately determined by the value of the feedback components. The frequency response of the amplifier is also improved, that is a wide range of frequencies are amplified by the same amount and the output of the amplifier, although increased or amplified, is a true copy of the input with little distortion.

Positive feedback

Positive feedback is the name given to the situation in which the feedback signal is added to the reference signal. This usually leads to instability and can be used in electronics to generate oscillations for use as the output of signal generators.

Instability

Instability can occur in any closed loop system because the output of each black box in the system forms the input to the next box. When a controlling action is required, the necessary changes can take some time to work their way through each block in the system from input to output. Let us suppose a reference sine wave was applied to one of the blocks in a system as shown in Figure 4.5. With negative feedback applied, the feedback signal will be negative with respect to the reference signal at any instant in time. The feedback signal

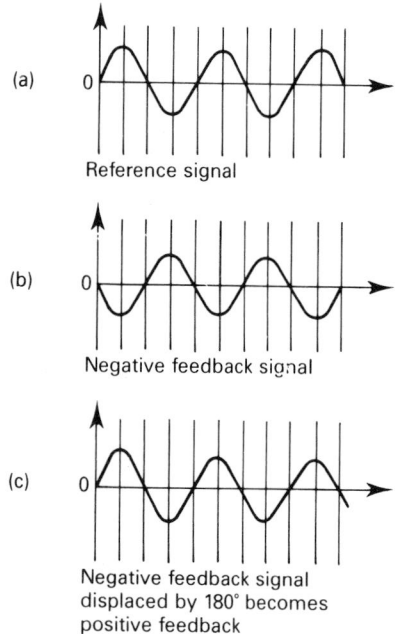

(a) Reference signal

(b) Negative feedback signal

(c) Negative feedback signal displaced by 180° becomes positive feedback

Figure 4.5 Effects of feedback

will, therefore, be subtracted from the reference signal, resulting in stability of the output. If there is a time lag in the system, resulting in a phase lag of say 180°, the feedback signal will be in-phase with the reference signal, resulting in positive feedback and instability of the system. An example of this effect is the howling of a loudspeaker when it is brought close to a microphone. The sound from the loudspeaker is picked up by the microphone but a time delay occurs as the sound waves travel through the air between the speaker and the microphone. This signal lag causes some frequencies to be amplified, resulting in the howl. Greater separation of speaker and microphone resolves this particular problem.

Transducers

A transducer is a device which converts a particular physical output into another type of physical output. The microphone and loudspeaker are both transducers, the microrophone converts sound waves into electrical signals and the loudspeaker converts electrical signals into sound waves. Other transducers are considered in Chapter 9. The output from the transducer is called the signal and the *signal conditioner* converts this output into a form which is compatible with the electrical circuits. The tacho-generator of Figure 4.3 converts speed of rotation into an electrical signal, for feedback to the operational amplifier acting as the error detector. In control systems the transducer often acts as a *sensor*, providing signals which allow a circuit to be controlled in a predetermined way. Other examples of sensors are solenoids, magnetic valves, pressure switches, microswitches, thermocouples and thermostats.

Washing machine control

A washing machine control system is made up of a number of sub-systems, many of which are sensors providing inputs to a management sub-system or controller. A typical system is shown in Figure 4.6. The water level sensor would typically be a small ball valve-operated microswitch which would switch off the water supply when a pre-determined level of water in the drum was reached. The water temperature sensor would be a thermostat switching off the electrical supply to the heater elements when a suitable washing temperature has been reached. The drum motor rotates the drum which contains the clothes and water, usually turning in one direction for a short time and then in the reverse direction to prevent the clothes becoming tangled. The pump is a water pump, removing waste water from the drum after each wash and rinse operation. The controller/timer is often a spring action rotary timer which incorporates a number of switch contacts or a semiconductor logic control device. This makes it possible to sequence the various operations such as fill the drum to the required level, raise the temperature of the water before rotating the drum and clothes. After a pre-determined time has elapsed, the pump removes the dirty water from the drum, the pump motor stops and clean water flows into the drum to the pre-determined level, before the drum is once more rotated for the rinsing cycle. Upon completion of the rinsing cycle the water is pumped out and the clothes are then ready for drying. All these operations need to be timed and many interconnections must be made. The controller/timer is the sub-system which manages these functions. If it becomes faulty the whole system will fail to operate. If the pump motor fails to pump, the drum will remain full of water. If the

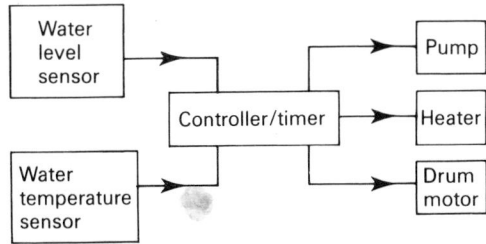

Figure 4.6 A washing machine control system

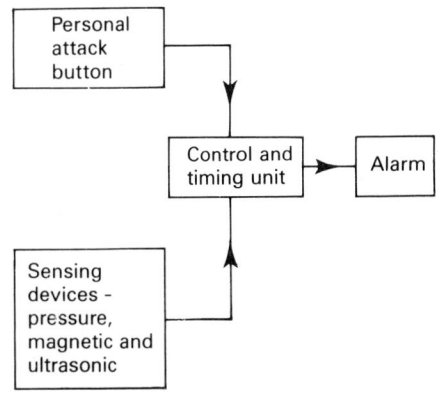

Figure 4.7 A security control system

Figure 4.8 A personal attack button

heater element fails, the clothes will be washed in water which is below the desired temperature. If the drum fails to rotate, the drum motor or drive belt may be broken and the clothes will simply soak in the water without being tossed about, which provides the best cleaning action. If the water temperature thermostat becomes open circuit, the water heater will not be switched on and the clothes will be washed in cold water. If the thermostat remains closed, the heater will remain switched on and the clothes will be washed in very hot water which may damage wool or delicate fabrics. If the water level sensor becomes faulty, the water will either not begin to fill or fill continuously until water overflows. The controller/timer permits some degree of sophisticated control by the interconnection of the various circuits and sensors. For example, the water heater will only operate when the correct level of water has been reached, and even then, only if the water is below the desired temperature.

Security systems

In recent years we have seen an increase in the installation of security systems to domestic and commercial premises for the protection of goods and property against the criminally inclined. The crime prevention authorities now all agree that the chance of a successful burglary is greatly reduced by fitting an effective alarm system. Figure 4.7 shows the system diagram of a typical security system. All burglar alarms consist of three subsystems, a warning device, detection devices and a control panel. The detection devices are the sensors of the system, which respond to the presence of an intruder or the opening of a window. The sensors trigger an audible warning device, usually an alarm bell and flashing light. The system is switched on and off at a control panel with a key or coded buttons. The control and timing unit is the master control panel for the whole security system. All the system components are joined together with small, low-voltage cable which is easily concealed.

The sensing devices, which detect the presence of a burglar, might be proximity switches on doors and windows or movement detectors such as passive infra red or ultrasonic detectors. A personal attack button is a press switch which is used to activate the alarm system manually in situations where the authorised occupants of the building become aware that an intruder is present. A personal attack button is shown in Figure 4.8.

Security system false alarms are most annoying to neighbours and the police. Water ingress can cause proximity switches to fail, movement sensors can be activated by domestic pets. However, by carefully selecting the equipment and adjusting the detector's sensitivity, false alarms can be minimised so that the system gives many years of trouble-free service and peace of mind. Security systems are further considered in Chapter 10.

Space heating control

A space heating system is today as much a part of the total installation in a domestic or commercial building as the electrical wiring or the plumbing and sanitary fittings. With the acceptance of central heating has grown the awareness of the convenience and cost effectiveness of controlling the system. In the 1950s a basic time clock was available to give simple on/off control but the technical advances made in the engineering of heating systems, particularly the introduction of flow valves and wholly pumped systems, has led to complete energy management systems in the 1990s. Programming and sensing units coupled to flow valves can give maximum economy and comfort for a selected temperature range at any time of the day or night. A typical gas-fired or oil-fired space heating system is shown in Figure 4.9.

The boiler provides the heat energy which is circulated around the system as a result of burning fuel, either oil or gas in most domestic systems. The fuel burner always incorporates a flame failure device such as a thermocouple, so that the fuel flow is switched off if the flame should fail for any reason.

The fuel control valve is an on/off valve controlling the supply of fuel to the boiler burner which is operated electromagnetically. When the programmer and hot water or radiators call for heat, the magnetic valve is energised, opening the valve and allowing fuel to flow to the burner. The fuel control valve is designed to fail safe and, therefore, a faulty valve would result in the boiler switching off. The programmer, room and water sensors would continue to call for heat but cold water only would be circulated because the boiler, the source of heat energy for the system, would have shut down as a result of the faulty valve.

Heat is transferred from the boiler, to the water heating cylinder or space heating radiators, by water contained in small bore pipes which are insulated to prevent heat losses. The water is transferred around the system by means of a water pump, usually a single phase induction motor which is energised when the programmer and hot water or radiators call for heat. If the pump fails, heat cannot be transferred to the hot water or radiators, but the boiler will ignite if the boiler thermostat is turned up when the programmer is switched on.

The controller can incorporate relays and connector blocks for the boiler, pump and valve switching; it is the management system responding to the input from the various sensors and activating the boiler fuel flow and pump. A three-way valve directs hot water from the boiler to, either the hot water cylinder, the space heating radiators or both, dependent upon the settings of the programmer, room and cylinder thermostats. If the controller fails, the link between the inputs and outputs will be broken and the system will be inoperative.

The room thermostat is the room temperature sensor. If the room temperature is below the thermostat setting, the circuit is closed, activating the flow valve and heat it transferred, by pumping hot water from the boiler to the space heating radiators. The room temperature rises until the thermostat setting is reached, the switch then opens, closing the flow valve and switching off the pump.

The cylinder thermostat is the water temperature sensor, usually clamped to the hot water storage cylinder. If the water temperature is below the thermostat setting, the circuit is closed, activating the flow valve, and heat is transferred by the pump from the boiler to the water contained in the cylinder. The domestic hot water temperature is raised until the thermostat setting is reached, the

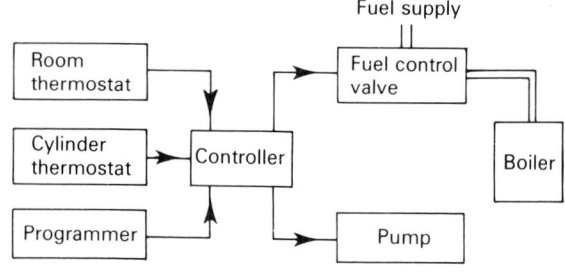

Figure 4.9 Space heating control system

switch then opens, closing the flow valve and switching off the pump.

The programmer incorporates a time clock to determine when the system is switched on or off and a selector switch which allows the user to predetermine what is required, space heating, hot water or both. With the programmer in the off mode, the whole system is shut down. With the programmer switched on, the system is designed to operate automatically, maintaining the water and room temperature at a predetermined setting through the flow valve, pump and temperature sensors.

If the programmer fails, the most likely cause will be a faulty time clock, which will result in the whole system being either switched off or on continuously.

Electronic communications

In electronics, as in many other engineering disciplines, we communicate many of our ideas and information with diagrams. Various types can be distinguished: block diagrams, wiring diagrams, circuit diagrams and supplementary diagrams. The one to use in a particular application is the one which most clearly communicates the desired information.

A block diagram

A block diagram is a relatively simple diagram in which an installation or piece of equipment is represented by block outlines. The purpose of a block diagram is to show clearly the operation or

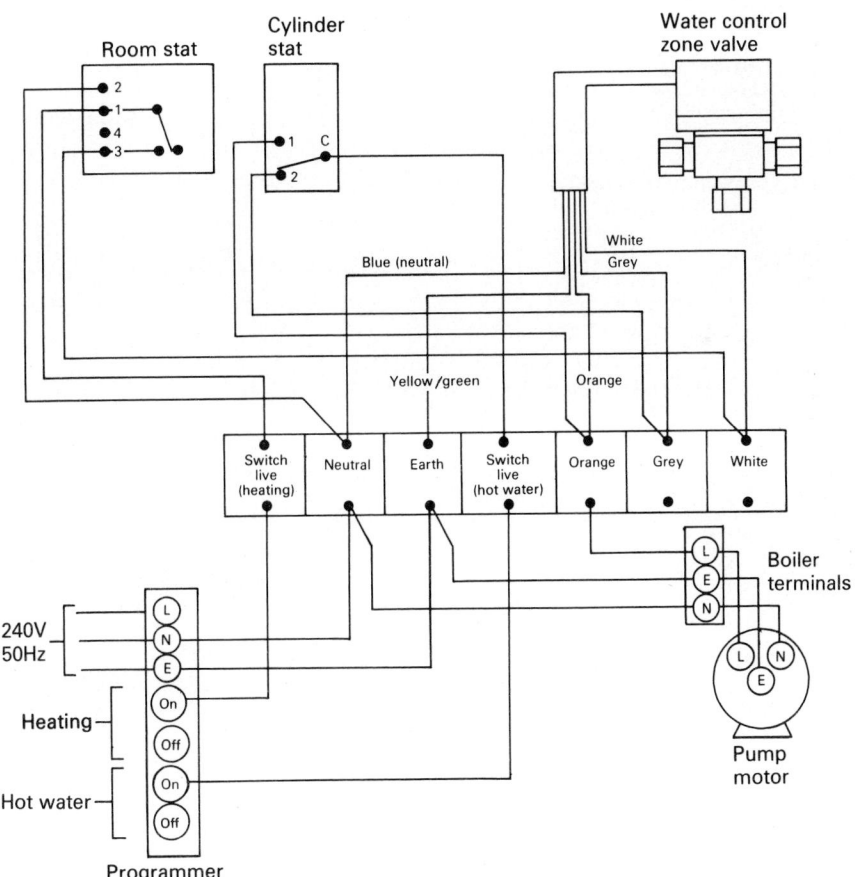

Figure 4.10 Wiring diagram for space heating control (Honeywell 'Y' plan)

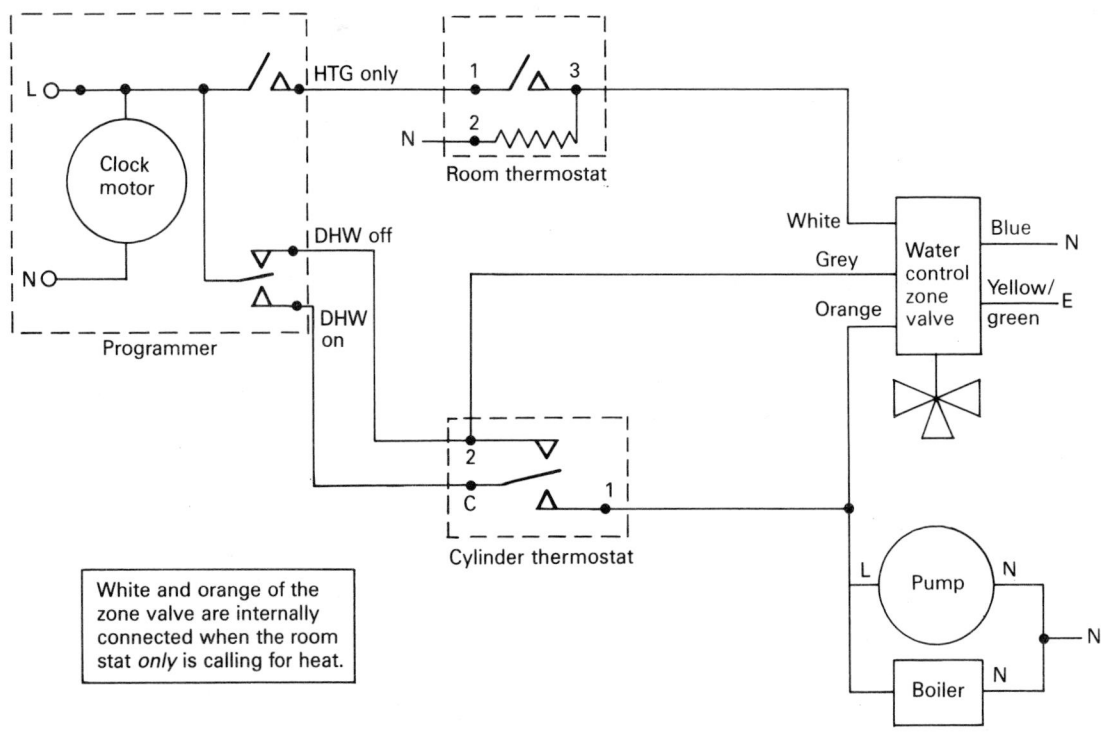

Figure 4.11 Circuit diagram for space heating control (Honeywell 'Y' Plan)

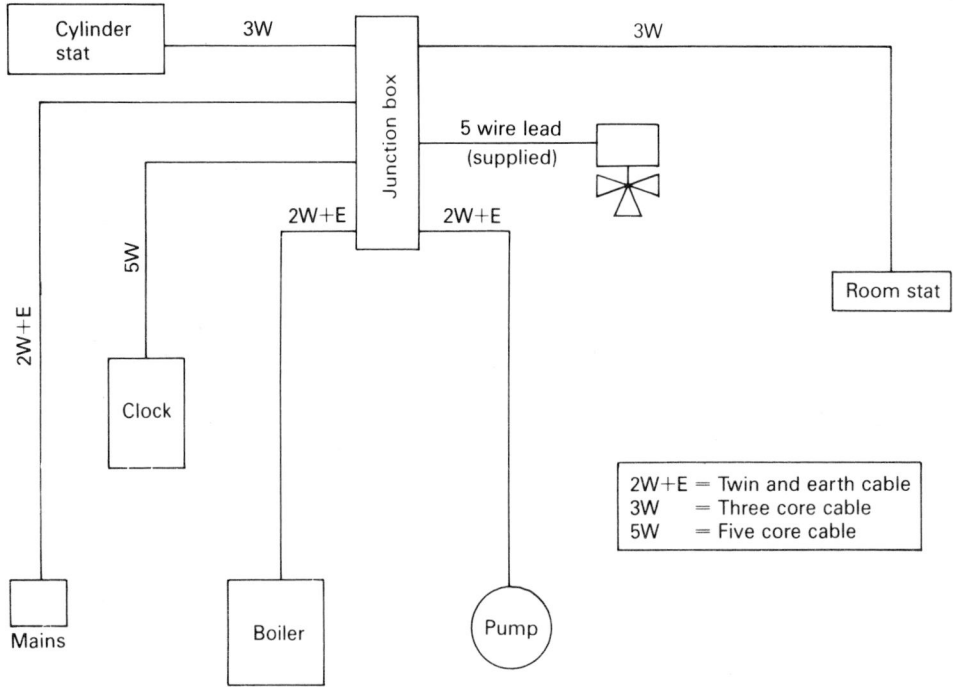

Figure 4.12 Supplementary diagram for a space heating system (Honeywell 'Y' plan)

function of a subject or system and, therefore, it does not normally show the physical layout of any components or the individual circuit connections. Figures 4.6, 4.7 and 4.9 are examples of block diagrams.

A wiring diagram

A wiring diagram or connection diagram shows the detailed connections between components or items of equipment and in some cases, the routing of these connections. The purpose of a wiring diagram is to help someone with the actual wiring of the circuit. Figure 4.10 is an example of a wiring diagram.

A circuit diagram

A circuit diagram shows most clearly how a circuit works. All the essential parts and connections are depicted by their graphical symbols. The purpose of a circuit diagram is to help in the understanding of the circuit and it should be laid out as clearly as possible without regard to the physical layout of the actual components or parts. Figure 4.11 is an example of a circuit diagram, as are most of the diagrams in Chapter 7.

A supplementary diagram

A supplementary diagram does not fall into any one of the above categories. Its purpose is to convey additional information in a form which is usually a mixture of the other three categories. Figure 4.12 shows a typical supplementary diagram, the cabling arrangements for a space heating system.

CHAPTER 5

Test equipment

The use of electronic circuits in electrical installation work has increased considerably over recent years. Electronic circuits and components can now be found in motor starting and control circuits, discharge lighting, emergency lighting, alarm circuits and special effects lighting systems. There is, therefore, a need for the installation electrician to become familiar with some basic electronic test equipment, which is the aim of this chapter.

Test instruments

Electrical installation circuits usually carry in excess of one ampere and often carry hundreds of amps. Electronic circuits operate in the milliampere or even micro-ampere range. The test instruments used on electronic circuits must have a *high impedance* so that they do not damage the circuit when connected to take readings. All instruments cause some disturbance when connected into a circuit because they consume some power in order to provide the torque required to move the pointer. In power applications these small disturbances seldom give rise to obvious errors, but in electronic circuits, a small disturbance can completely invalidate any readings taken. We must, therefore, choose our electronic test equipment with great care. Let us consider some of the problems.

Let me first of all define what is meant by the terms error and accuracy used in this chapter. When the term *error* is used it means the *deviation of the meter reading from the true value* and *accuracy* means the *closeness of the meter reading to the true value*.

Instrument errors

Consider a voltmeter of resistance $100\,k\Omega$ connected across the circuit shown in Figure 5.1(a).

Connection of the meter loads the circuit by effectively connecting a $100\,k\Omega$ resistor in parallel with the circuit resistor as shown in Figure 5.1(b) which changes the circuit to that shown in Figure 5.1(c).

Common sense tells us that the voltage across each resistor will be 100 volts but the meter would read about 66 volts because connection of the meter has changed the circuit. This loading effect can be reduced by choosing instruments which have a very high impedance. Such an instrument imposes less load on the circuit and gives an indication much closer to the true value.

The deflection torque of most instruments is proportional to current and since current $I = V/Z$ and $Z^2 = R^2 + X_L^2$ and $X_L = 2\pi fL$ the instrument is also frequency dependent. The important practical consideration is the *frequency range* of the test instrument. This is the range of frequencies over which the instrument may be considered free from frequency errors and is indicated on the back of the instrument or in the manufacturer's information. Frequency limitations are not a normal consideration for an electrician since electrical installations operate at the fixed mains frequency of 50 Hz.

The scale calibration of an instrument assumes a sinusoidal supply unless otherwise stated. Non-sinusoidal or complex waveforms contain harmonic frequencies which may be outside the instrument frequency range. The chosen instrument must, therefore, be suitable for the test circuit waveform.

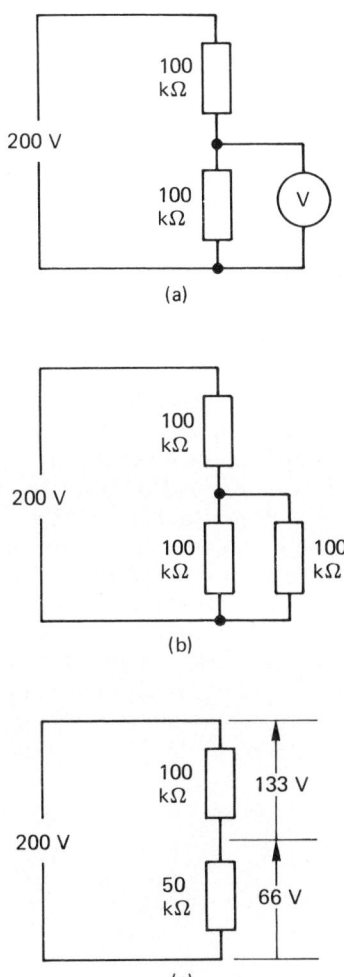

Figure 5.1 Circuit disturbances caused by the connection of a voltmeter

The maximum permissible errors for various instruments and their applications are indicated in British Standard 89. When choosing an instrument for electronic testing an electrician will probably be looking for an instrument with about a 2% maximum error, that is 98% accurate. Instrument manufacturers will provide detailed information for their products.

Errors are not only restricted to the instrument being used, operators can cause errors too.

Operator errors

These are errors such as mis-reading the scale, reading 28.3 and tabulating 23.8 or reading the wrong scale on a multi-range instrument. The test instrument must be used on the most appropriate scale, don't try to read 12 V on a 250 V scale, the reading will be much more accurate if the 25 V scale is used to read a value of about 12 V.

The type of instrument to be purchased for general use is a difficult choice because there are so many different types on the market and every manufacturer's representative is convinced that his company's product is the best. However, most instruments can be broadly grouped under two general headings: those having *analogue* or *digital* displays.

Analogue and digital displays

Analogue meters

These meters have a pointer moving across a calibrated scale. They are the only choice when a general trend or variation in value is to be observed. Hi-fi equipment often uses analogue displays to indicate how power levels vary with time, which is more informative than a specific value. Red or danger zones can be indicated on industrial instruments. The fuel gauge on a motor car often indicates full, half full or danger on an analogue display which is much more informative than an indication of the exact number of litres of petrol remaining in the tank.

These meters are only accurate when used in the calibrated position – usually horizontally.

Most meters using an analogue scale incorporate a mirror to eliminate parallax error. The user must look straight at the pointer on the scale when taking readings and the correct position is indicated when the pointer image in the mirror is hidden behind the actual pointer. A good-quality analogue multimeter suitable for electronic testing is shown in Figure 5.4.

The input impedance of this type of instrument is typically 1000 Ω per volt or 20,000 Ω per volt, depending upon the scale chosen.

Digital meters

These provide the same functions as analogue meters but they display the indicated value using a

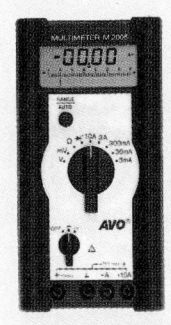

Figure 5.2 Digital multimeter suitable for testing electronic circuits

seven segment LED (see Figure 6.11) to give a numerical value of the measurement. Modern digital meters use semiconductor technology to give the instrument a very high input impedance, typically about $10\,M\Omega$ and, therefore, they are ideal for testing most electronic circuits.

The choice between a meter having an analogue or digital display is a difficult one and must be dictated by specific circumstances. However, if you are an electrician intending to purchase a new instrument which would be suitable for electronic testing, I think on balance that a good-quality digital multimeter such as that shown in Figure 5.2 would be best. Having no moving parts, digital meters tend to be more rugged and, having a very high input impedance, they are ideally suited to testing electronic circuits.

The multimeter

Multimeters are designed to measure voltage, current or resistance. Before taking measurements the appropriate volt, amp or ohm scale should be selected. To avoid damaging the instrument it is good practice first to switch to the highest value on a particular scale range. For example, if the 10 A scale is first selected and a reading of 2.5 A is displayed, we then know that a more appropriate scale would be the 3 A or 5 A range. This will give a more accurate reading which might be, say, 2.49 A. When the multimeter is used as an ammeter to measure current it must be connected in series with the test circuit as shown in Figure 5.3. When used as a voltmeter the multimeter must be connected in parallel with the component as shown in Figure 5.3.

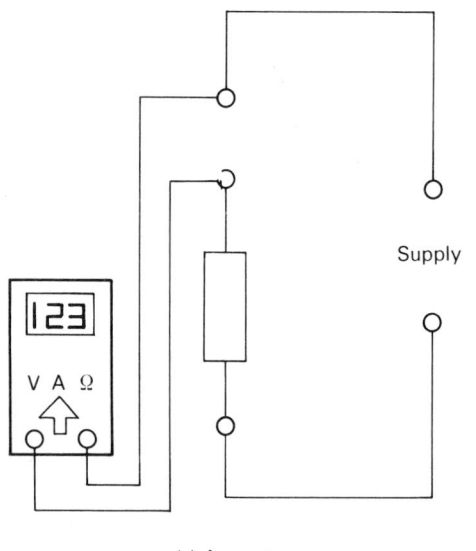

(a) Ammeter

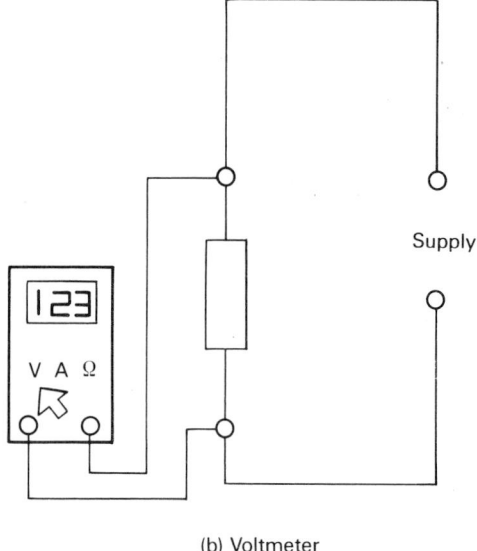

(b) Voltmeter

Figure 5.3 Using a multimeter (a) as an ammeter and (b) as a voltmeter

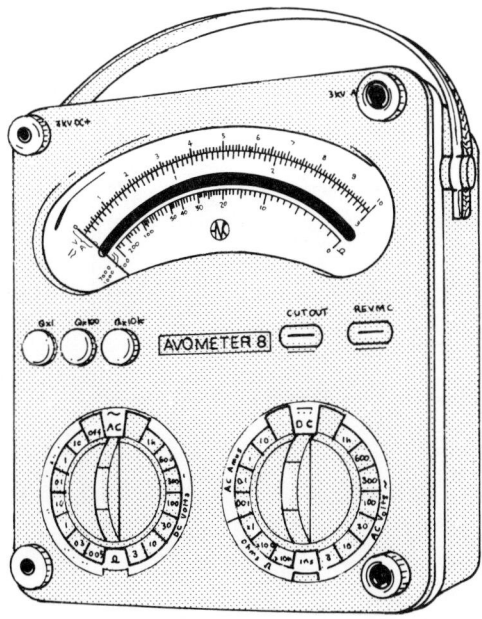

Commonly used multirange instrument

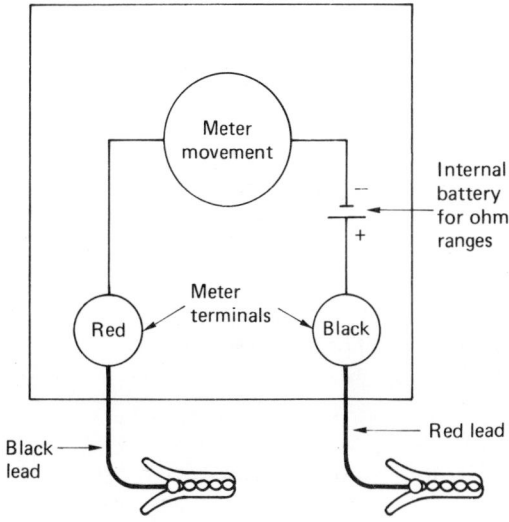

Figure 5.4 Multirange meter used as an ohm meter

Ohm meter

When using a commercial multi-range meter as an ohm meter for testing electronic components, care must be exercised in identifying the positive terminal. The red terminal of the meter, identifying the positive input for testing voltage and current, usually becomes the negative terminal when used as an ohm meter because of the way the internal battery is connected to the meter movement. To reduce confusion when using a multi-range meter as an ohm meter it is advisable to connect the red lead to the black terminal and the black lead to the red terminal so that the red lead indicates positive and the black lead negative as shown in Figure 5.4. The ohm meter can then be successfully used to test diodes, transistors and thyristors as described in Chapter 6, and resistors and capacitors as described in Chapter 1.

Commercial multi-range instruments reading volts, amps and ohms are usually the most convenient test instrument for an electrician, although a cathode ray oscilloscope (CRO) can be invaluable for bench work.

The cathode ray oscilloscope (CRO)

The CRO is probably one of the most familar and useful instruments to be found in an electronic repair service workshop or college laboratory. It is a most useful instrument for two reasons, it is a high impedance voltmeter and, therefore, takes very little current from the test circuit and, secondly, it allows us to 'look into' a circuit and 'see' the waveforms present. The cathode ray is the name given to a high-speed beam of electrons generated in the cathode ray tube and was first used during the Second World War as part of the *Radar* system. The beam of electrons is deflected horizontally across the screen at a constant rate by the *time-base circuit* and vertically by the test voltage. The many controls on the front of the CRO are designed so that the operator can stabilise and control these signals. Figure 5.5 shows the front panel of a simple CRO. Electricians who are unfamiliar with the CRO should not be baffled by the formidable array of knobs and switches – take

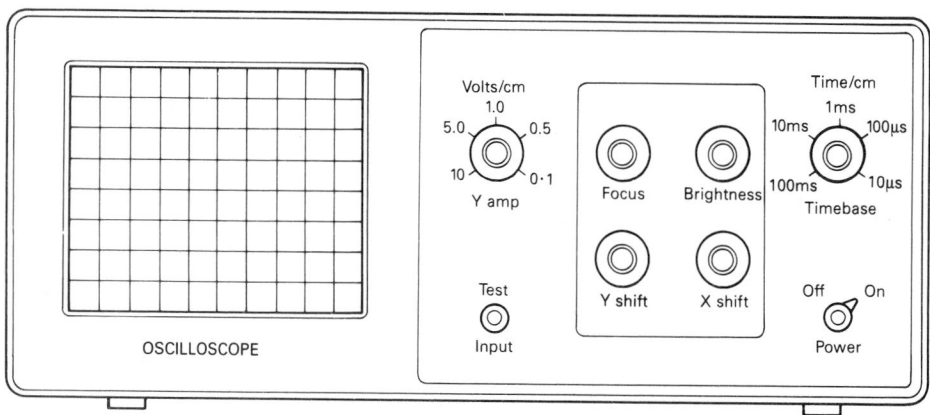

Figure 5.5 Front panel of a simple CRO

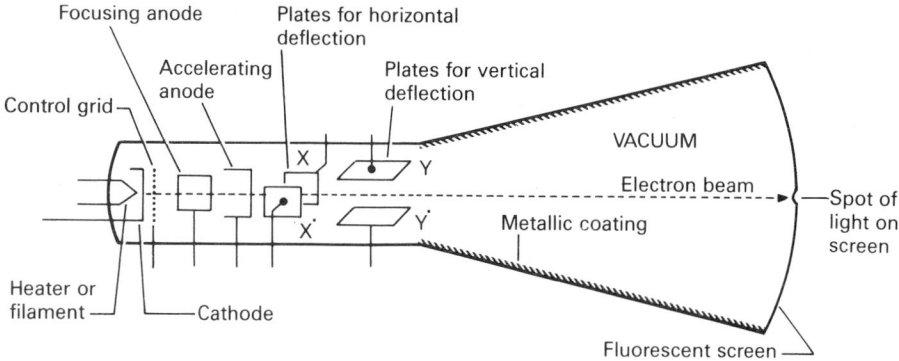

Figure 5.6 Simplified diagram of the cathode ray tube

them one at a time – and give yourself time to become familiar with these controls.

The single most important component in the CRO is the cathode ray tube.

Cathode ray tube

Figure 5.6 shows a simplified diagram of the cathode ray tube. This is an evacuated glass tube containing the *electron gun* components on the left and the fluorescent screen, which the operator looks at, on the right. On the far left of the diagram is the wire filament through which a current is passed. This heats a metal plate called a cathode,

which emits the electrons to be accelerated. The *rate* at which the electrons are accepted for acceleration could be modified by making changes to the temperature of the cathode, but in practice it is more convenient to have a metal control grid with a hole in it. By varying the voltage of the control grid it is possible to influence the number of electrons passing through the hole in the grid. The electrons which pass through the grid tend to be moving in various directions and the purpose of the next component therefore is to focus the beam. The electrons are then further accelerated by the accelerating anode to give them sufficient final velocity to produce a bright spot on the screen.

The electrons, on emerging through a hole in this anode, pass through two pairs of parallel plates X X' and Y Y', each pair being at right-angles to the other. If an electric field is established between X and X' the beam can be deviated horizontally, the direction and magnitude of the deflection depending upon the polarity of the plates. The negative beam of electrons is attracted towards the more positive plate. Likewise, an electric field between plates Y and Y' produces a vertical deviation. Therefore, a suitable combination of electric fields across X X' and Y Y' directs the beam to any desired point on the screen.

Upon reaching the screen, the electrons bombard the fluorescent coating on the inside of the screen and emit visible light. The brightness of the spot depends upon the speed of the electrons and the number of electrons arriving at that point.

Use of the CRO

The function of the various controls is outlined as follows:

1. Power on switch – Switch on and wait a few seconds for the instrument to warm up. An LED usually indicates a satisfactory main supply.
2. Brightness or Intensity – This controls the brightness of the trace. This should be adjusted until bright, but not over brilliant, otherwise the fluorescent powder may be damaged.
3. Scale illumination – This illuminates and highlights the 1 cm square grid lines on the screen.
4. Focus – The spot or trace should be adjusted for a sharp image.
5. Gain controls – Adjust for calibrate.
6. X-shift – The spot or trace can be moved to the left or right and should be centralised.
7. Y-shift – The spot or trace can be moved up or down.
8. TRIG control – This allows the time base to be synchronised to the applied signal to enable a steady trace to be obtained. Set the switch to either Auto or to the Y input which is connected to the test voltage.
9. AC/GND/DC – It is quite common for a signal to be made up of a mixture of AC and DC. Select DC for all signals and AC to block out the d.c. component of a.c. signals. The GND position disconnects the signal from the Y amplifier and connects the Y plates to ground or earth.
10. Chop/Alt – When a double beam oscillo-scope is used, it is common practice to obtain the two X traces from one beam by either sweeping the electron beams alternately or by sweeping a very small segment of each beam as the trace moves across the screen, leaving each trace chopped up. Use *chop* for slow time base ranges and *Alt* for fast time base ranges.
11. Connect – Connect the test voltage to the CRO leads and adjust the calibrated Y-shift (volts/cm) and time base (time/cm) controls until a steady trace fills the screen.

Use of the CRO to measure voltage and frequency

The calibrated Y-shift, time base and 1 cm grating on the tube front provide us with a method of measuring the displayed waveform.

With the test voltage connected to the Y-input, adjust all controls to the calibrate position. Adjust the X and Y tuning controls until a steady trace is obtained on the CRO screen, such as that shown in Figure 5.7.

To measure the voltage of the signal shown in Figure 5.7 count the number of centimetres from one peak of the waveform to the other using the centimetre grating. This distance is shown as 4 cm in Figure 5.7. This value is then multiplied by the volts/cm indicated on the Y amplifier control knob. If the knob was set to, say, 2 V/cm, the

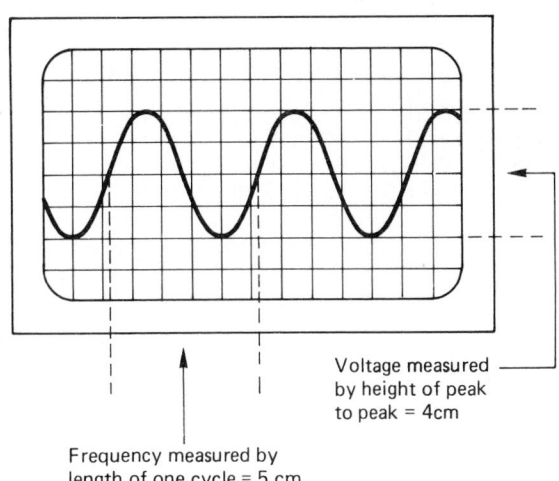

Voltage measured by height of peak to peak = 4cm

Frequency measured by length of one cycle = 5 cm

Figure 5.7 Typical trace on a CRO screen

peak-to-peak voltage of Figure 5.7 would be 4 cm × 2 V/cm = 8 V. The peak voltage would be 4 V and the rms voltage 0.7071 × 4 = 2.828 V.

To measure the frequency of the waveform shown in Figure 5.7 count the number of centimetres for one complete cycle using the one-centimetre grating. The distance is shown as 5 cm in Figure 5.7. This value is then multiplied by the time/cm on the X amplifier or time base amplifier control knob. If this knob was set to 4 ms/cm the time taken to complete one cycle would be 5 cm × 4 ms/cm = 20 ms. Frequency can be found from:

$$f = \frac{1}{T} \text{ (Hz)}$$

$$\therefore\ f = \frac{1}{20 \times 10^{-3}} = \frac{1000}{20} = 50 \text{ Hz}.$$

The waveform shown in Figure 5.7 therefore has an rms voltage of 2.828 V at a frequency of 50 Hz. The voltage and frequency of any waveform can be found in this way. The relevant a.c. theory is covered in Chapter 3.

Example 1

A sinusoidal waveform is displayed on the screen of a CRO as shown in Figure 5.7. The controls on the Y axis are set to 10 V/cm and the measurement from peak to peak is measured as 4 cm. Calculate the rms value of the waveform.

The peak-to-peak voltage is 4 cm × 10 V/cm = 40 V
The peak voltage is 20 V.
The rms voltage is 20 V × 0.7071 = 14.14 V.

Example 2

A sinusoidal waveform is dispalyed on the screen of a CRO as shown in Figure 5.7. The controls on the X axis are set to 2 ms/cm and the measurement for one period is calculated to be 5 cm. Calculate the frequency of the waveform.

The time taken to complete one cycle (T) is 5 cm × 2 ms/cm = 10 ms.

$$\text{Frequency} = \frac{1}{T} \text{ (Hz)}$$

$$\therefore\ f = \frac{1}{10 \times 10^{-3}} = \frac{1000}{10} = 100 \text{ Hz}.$$

As you can see, the CRO can be used to calculate the values of voltage and frequency. It is not a *direct reading* instrument as were the analogue and digital instruments considered previously. It does, however, allow us to observe the quantity being measured unlike any other instrument and, therefore, makes a most important contribution to our understanding of electronic circuits.

Signal generators

A signal generator is an oscillator which produces an a.c. voltage of continuously variable frequency. It is used for serious electronic testing, fault-finding and experimental work. One application for a signal generator is to test the frequency response of an audio amplifier to a range of frequency.

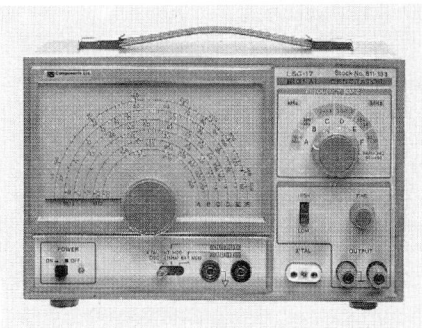

Figure 5.8 A signal generator

The human ear has a frequency range of approximately 15 Hz to 15 kHz and, therefore, an audio amplifier must respond to at least this range of frequencies. This test is described in detail in Chapter 7 'Testing audio amplifiers'. A signal generator is shown in Figure 5.8.

Power supply unit (P.S.U.)

A bench power supply unit is a very convenient way of obtaining a variable d.c. voltage from the a.c. mains. The output is very pure, a straight line when observed on a CRO, and continuously variable from zero to usually 30 V. It provides a convenient power source for bench testing or building electronic circuits. A bench power supply unit is shown in Figure 5.9.

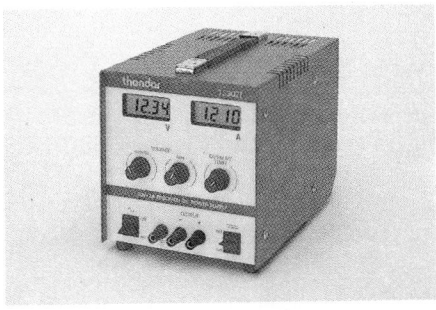

Figure 5.9 A bench power supply unit (PSU)

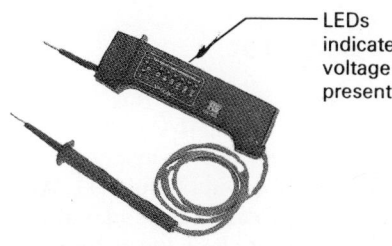

LEDs indicate voltage present

Figure 5.10 A voltage indicator

Mains electricity supply

The mains electricity supply can be lethal as all electricians will know. It is, therefore, a sensible precaution to connect any electronic equipment being tested or repaired to a socket protected by a residual current device. Electronic equipment is protected by in-line fuses and circuit breakers and when testing suspected faulty electronic equipment, a good starting-point is to establish the presence of the mains supply. A multi-range meter with the 250 V range selected would be a suitable instrument for this purpose or, alternatively, a voltage indicator as shown in Figure 5.10 could be used.

When isolating electronic equipment from the mains supply, in order to carry out tests or repairs, the following procedure should be followed:

1. Connect the voltage indicator or voltmeter to the incoming supply of the piece of equipment to be isolated. This should indicate the mains voltage and proves the effectiveness of the test instrument.
2. Isolate the supply.
3. Again test the supply to the equipment. If zero volt is indicated the equipment is disconnected from the mains supply and safe to work on.

Insulation tester

The use of an insulation resistance test as described by Regulation 613-5 of the IEE Regulations must be avoided with any electronic equipment. The working voltage of this instrument can cause total devastation to modern electronic equipment. When carrying out an insulation resistance test as part of the prescribed series of tests for an electrical installation, all electronic equipment must first be disconnected or damage will result.

Any resistance measurements made on electronic circuits must be achieved with a battery-operated ohm meter as described previously to avoid damaging the electronic components.

Measurement of speed of rotation

In most methods of measuring speed of rotation the instrument is coupled to the shaft whose speed is to be measured. The coupling may be direct with the instrument permanently connected or the instrument may be hand held to the end of the shaft while the measurement is being made.

The tachometer

This is usually a hand-held instrument on which speed is registered directly on a calibrated dial in revs/sec. Tachometers are usually operated mechanically, they are useful for spot checks of speed but accuracy is not very good.

The tachogenerator

This is a permanently coupled generator connected to the shaft end which produces a voltage proportional to the shaft speed. This allows a voltmeter, suitably calibrated in revs/sec, to be placed at some distance from the machine whose speed is to be measured.

The stroboscope

The stroboscope measures rotational speed by 'freezing' the movement so that the rotating body appears stationary. If a mark is made on the rotating shaft with chalk or whitener and a light flashed every time the mark is in the same position, the eye would only see the mark in that one position, and, therefore, the shaft would appear stationary. When taking measurements with a stroboscope the instrument is pointed at the marked shaft and the rate at which the lamp flashes is varied until the shaft appears stationary. The rate of flashing is then a measure of speed and the control dial is usually calibrated directly in revs/sec.

The advantage of this method of speed measurement is that there is no physical connection between the rotating shaft and the test equipment. The speed of very small motors found in control systems can only be measured in this way because the power required to drive a tachometer, for example, would cause the motor under investigation to slow down.

The disadvantage of this method is that we also see stationary images if the mark is illuminated once every two or three revolutions and so on. Readings are, therefore, obtained for a half, third or a quarter of the true speed. One way to avoid this error is by using some other speed-measuring

device as an initial check or by calculating the approximate speed. Good practice suggests that the stroboscope should be used as follows: With the stroboscope aimed at the marked shaft, its flash rate is slowly and steadily reduced, starting with the highest rate and working down the speed range until the stationary mark is observed.

The stroboscope principle of 'freezing' movement makes this equipment useful for examining the behaviour of industrial processes which are moving too rapidly to be seen normally with the human eye. Such effects as vibrating objects, valve bounce or the turbulence of the air flow over fan blades can be studied in detail while the parts are moving at high speed.

Pulse counter

The three components of this system are a 60-toothed wheel attached to the shaft whose speed is to be measured, an electro-magnet and an electronic counter as shown in Figure 5.11. As each tooth of the wheel approaches the end of the bar magnet the steel tooth provides a lower resistance magnetic path than did the gap between the teeth and this increases the flux density in the electro-magnet and induces a voltage with a frequency equal to the number of teeth passing the electro-magnet per second. The pulse counter amplifies the signal and

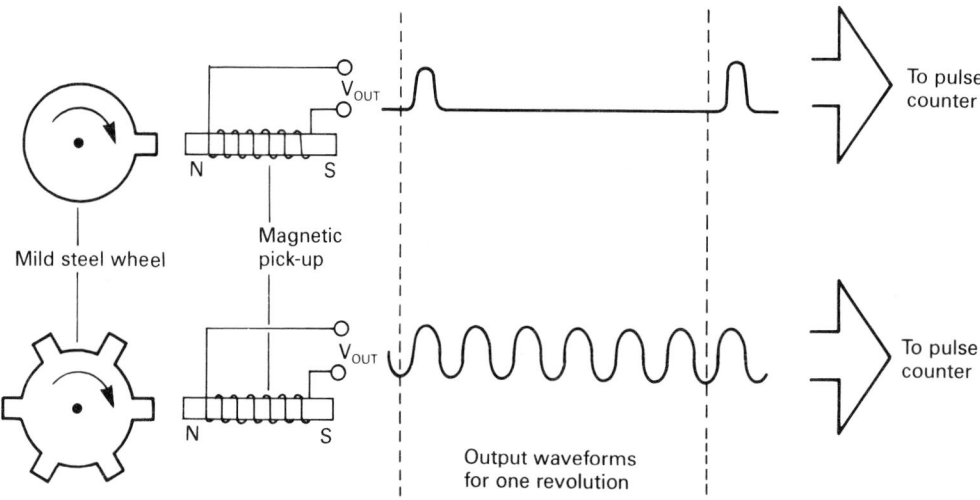

Figure 5.11 Measuring speed of rotation by counting magnetic pulses

displays it on an LED. The display can be mounted at some distance from the machine whose speed is being measured. Connection must be made by screened co-axial cables to prevent additional pulses being picked up from adjacent electrical equipment such as contactors or automatic star-delta motor starters.

Figure 5.12 shows a miniature magnetic pick-up transducer which incorporates a semiconductor I.C. to give a digital output every time ferrous metal passes the pole piece. The output waveform is compatible with most logic systems and is shown

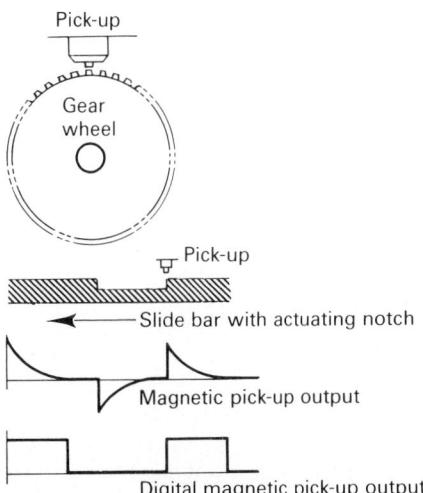

Figure 5.12 Miniature magnetic pick-up transducer

Figure 5.13 Output from a miniature magnetic pick-up

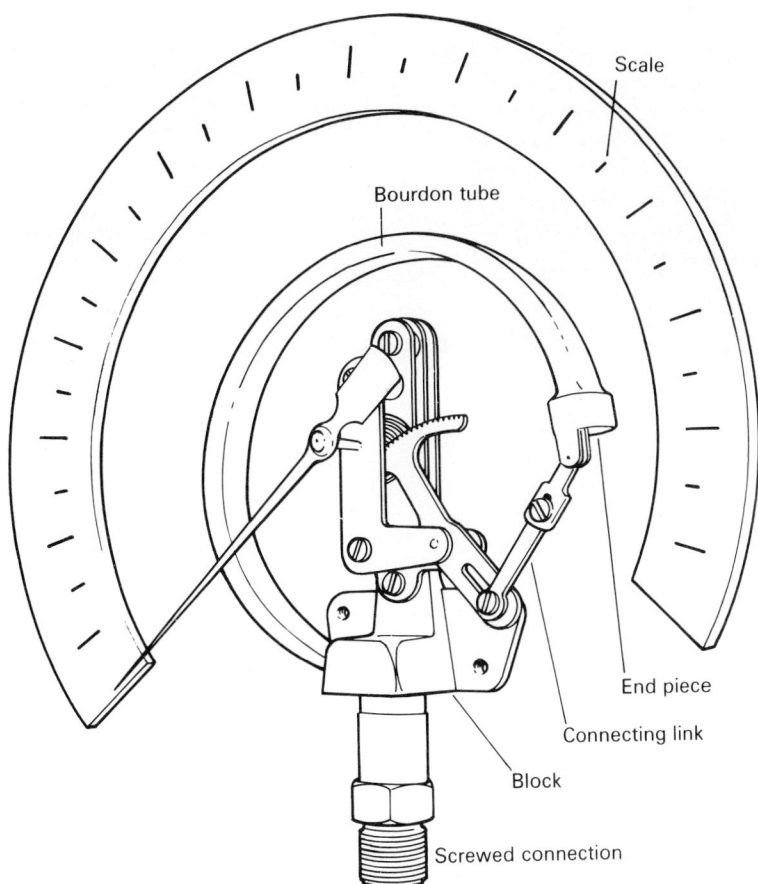

Figure 5.14 A Bourdon tube pressure gauge

in Figure 5.13. The pick-up is encased in a screwed steel cylinder 30 mm long and 6.35 mm in diameter which makes it small enough to be deployed where many conventional sensors could not be fitted.

The Bourdon tube pressure gauge

The standard type of pressure gauge is shown in Figure 5.14 and consists of a Bourdon tube driving a pointer through gearing across a scale. The Bourdon tube has an oval cross-section, closed at one end and bent into the arc of a circle, as shown in Figure 5.15. When pressure is admitted, the tube tries to straighten out so that it becomes an arc of greater radius. The change is very small but the toothed quadrant and pinion mechanically amplify the displacement and the pointer moves across a suitably calibrated scale.

The electrical measurement of non-electrical quantities is further discussed in Chapter 9 of this book. The principle and operation of a number of electrical measuring instruments are covered in Chapter 8 of *Advanced Electrical Installation Work*. The detailed inspection and testing procedures required by the IEE Regulations for Electrical Installations are covered in Chapter 9 of *Basic Electrical Installation Work*.

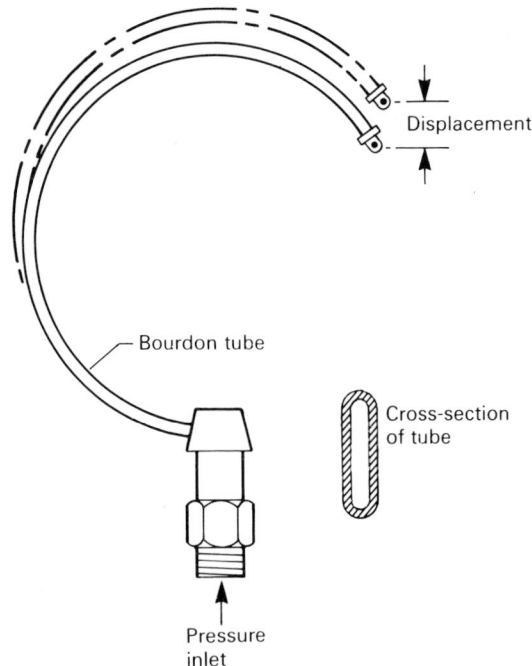

Figure 5.15 Operation of the Bourdon tube

CHAPTER 6

Semiconductor devices

Semiconductor materials

Modern electronic devices use the semiconductor properties of materials such as silicon or germanium. The atoms of pure silicon or germanium are arranged in a lattice structure as shown in Figure 6.1. The outer electron orbits contain four electrons known as valence electrons. These electrons are all linked to other valence electrons from adjacent atoms forming a co-valent bond. There are no free electrons in pure silicon or germanium and, therefore, no conduction can take place unless the bonds are broken and the lattice framework is destroyed.

To make conduction possible without destroying the crystal it is necessary to replace a four-valent atom with a three- or five-valent atom. This process is known as *doping*.

If a three-valent atom is added to silicon or germanium a hole is left in the lattice framework. Since the material has lost a negative charge, the material becomes positive and is known as a p-type material, p for positive.

If a five-valent atom is added to silicon or germanium, only four of the valence electrons can form a bond and one electron becomes mobile or free to carry charge. Since the material has gained a negative charge it is known as an n-type material, n for negative.

Bringing together a p-type and n-type material allows current to flow in one direction only through the pn junction. Such a junction is called a diode since it is the semiconductor equivalent of the vacuum diode valve used by Fleming to rectify radio signals in 1904.

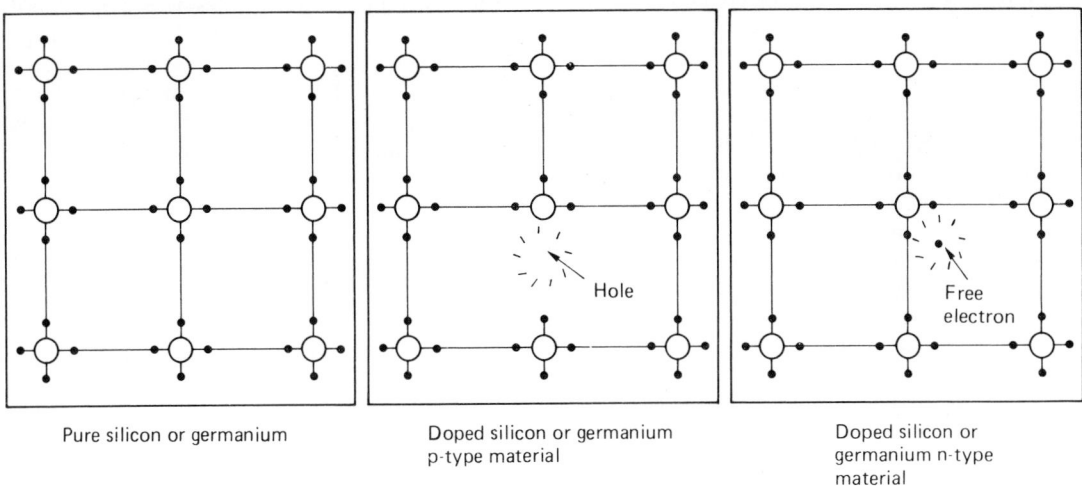

Pure silicon or germanium

Doped silicon or germanium p-type material

Doped silicon or germanium n-type material

Figure 6.1 Semiconductor material

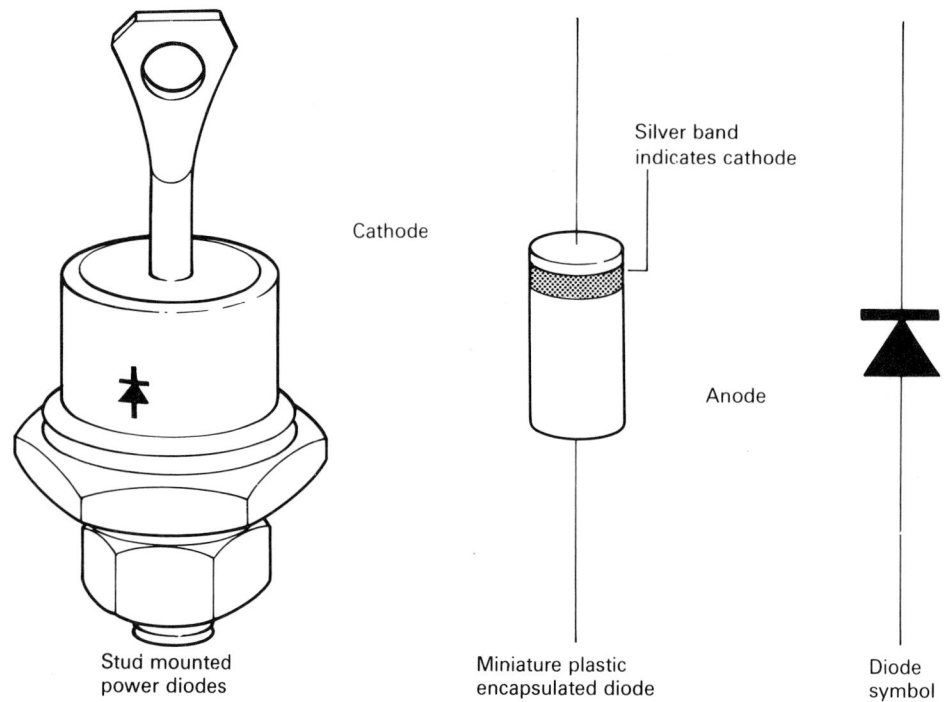

Cathode

Silver band
indicates cathode

Anode

Stud mounted
power diodes

Miniature plastic
encapsulated diode

Diode
symbol

Figure 6.2 Symbol and appearance of semiconductor diodes

Semiconductor diode

A semiconductor or junction diode consists of a p-type and n-type material formed in the same piece of silicon or germanium. The p-type material forms the anode and the n-type the cathode as shown in Figure 6.2. If the anode is made positive with respect to the cathode, the junction will have very little resistance and current will flow. This is referred to as forward bias. However, if reverse bias is applied, that is, the anode is made negative with respect to the cathode, the junction resistance is high and no current can flow as shown in Figure 6.3. The characteristics for a forward and reverse bias p-n junction are given in Figure 6.4.

It can be seen that a small voltage is required to forward bias the junction before a current can flow. This is approximately 0.6 V for silicon and 0.2 V for germanium. The reverse bias potential of silicon is about 1200 V and for germanium about 300 V. If the reverse bias voltage is exceeded the diode will break down and current will flow in both directions. Similarly, the diode will break down if

the current rating is exceeded, because excessive heat will be generated. Manufacturers' information therefore gives maximum voltage and current ratings for individual diodes which must not be exceeded (see Appendix F). However, it is possible to connect a number of standard diodes in series or parallel, thereby sharing current or voltage as shown in Figure 6.5, so that the manufacturers' maximum values are not exceeded by the circuit.

Diode testing

The p-n junction of the diode has a low resistance in one direction and a very high resistance in the reverse direction.

Connecting an ohm meter, as described in Chapter 5, with the red positive lead to the anode of the junction diode and the black negative lead to the cathode, would give a very low reading. Reversing the lead connections would give a high resistance reading in a 'good' component.

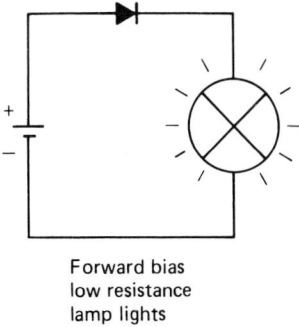

Forward bias
low resistance
lamp lights

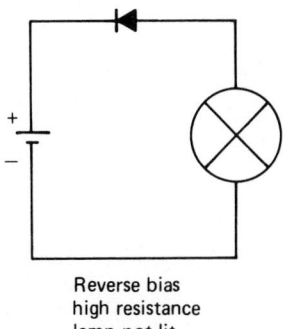

Reverse bias
high resistance
lamp *not* lit

Figure 6.3 Forward and reverse bias of a diode

Zener diode

A zener diode is a silicon junction diode but with a different characteristic to the semiconductor diode considered previously. It is a special diode with a pre-determined reverse breakdown voltage, the mechanism for which was discovered by Carl Zener in 1934. The symbol and general appearance is shown in Figure 6.6. In its forward bias mode, that is when the anode is positive and the cathode negative, the zener will conduct at about 0.6 V, just like an ordinary diode but it is in the reverse mode that the zener diode is normally used. When connected with the anode made negative and the cathode positive, the reverse current is zero until the reverse voltage reaches a pre-determined value when the diode switches on as shown by the characteristics given in Figure 6.7. This is called the zener voltage or reference voltage. Zener diodes are manufactured in a range of preferred values, for example, 2.7 V, 4.7 V, 5.1 V, 6.2 V, 6.8 V, 9.1 V, 10 V, 11 V, 12 V etc., up to 200 V at various ratings, as shown by the tables in the appendix. The diode may be damaged by overheating if the current is not limited by a series resistor, but when this is connected, the voltage across the diode remains constant. It is this

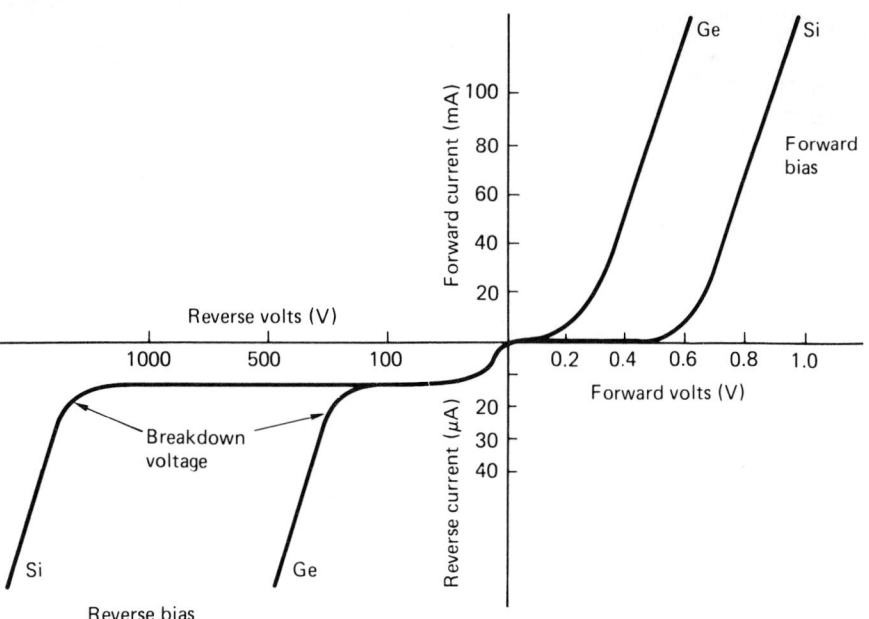

Figure 6.4 Forward and reverse characteristic of silicon and germanium

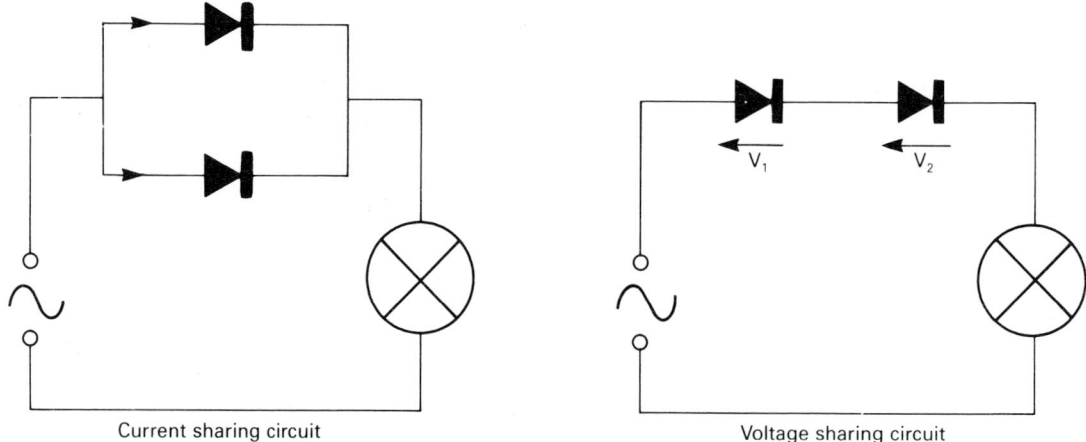

Current sharing circuit   Voltage sharing circuit

Figure 6.5 Using two diodes to reduce the current or voltage applied to a diode

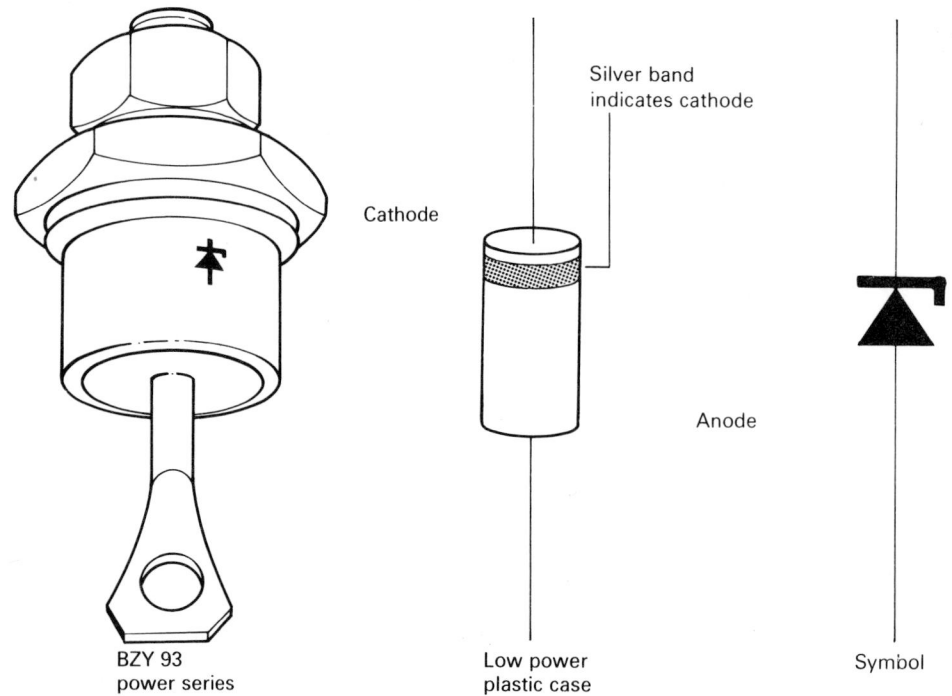

Figure 6.6 Symbol and appearance of Zener diodes

property of the zener diode which makes it useful for stabilising power supplies and these circuits are considered in Chapter 7.

If a test circuit is constructed as shown in Figure 6.8, the zener action can be observed. When the supply is less than the zener voltage (5.1 V in this case) no current will flow and the output voltage will be equal to the input voltage. When the supply is equal to or greater than the zener voltage, the diode will conduct and any excess voltage will appear across the 680 Ω resistor resulting in a very stable voltage at the output. When connecting this and other electronic circuits you must take care to connect the polarity of the zener diode as shown in the diagram. Note that current must flow through the diode to enable it to stabilise.

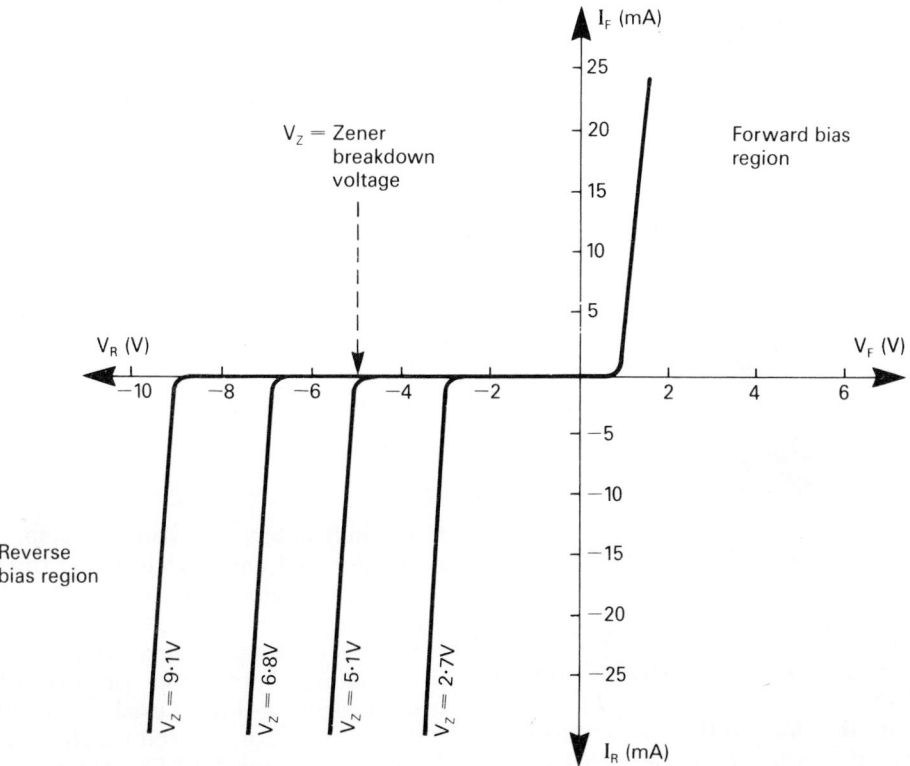

Figure 6.7 Zener diode characteristics

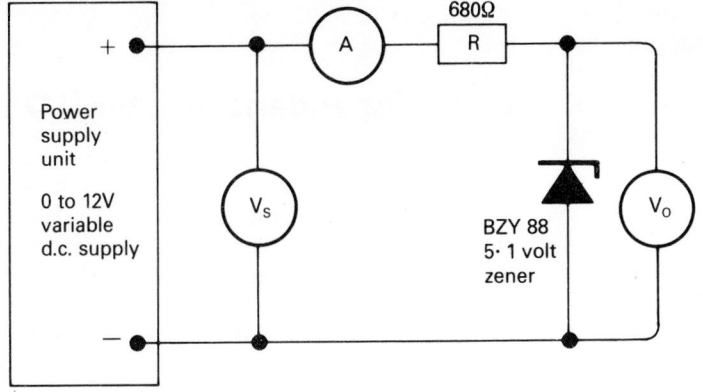

P.S.U. Supply Volts V_S	Current A	Output Volts V_O
1		
2		
3		
4		
5		
6		
7		
8		
9		
10		
11		
12		

Figure 6.8 Experiment to demonstrate the operation of a Zener diode

Light-emitting diode (LED)

The light-emitting diode is a p-n junction especially manufactured from a semiconducting material which emits light when a current of about 10 mA flows through the junction.

No light is emitted when the junction is reverse biased and if this exceeds about 5 V the LED may be damaged.

The general appearance and circuit symbol are shown in Figure 6.9.

The LED will emit light if the voltage across it is

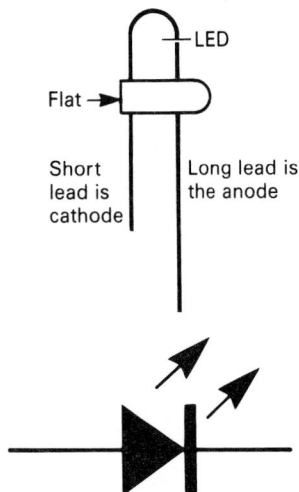

Figure 6.9 Symbol and general appearance of an LED

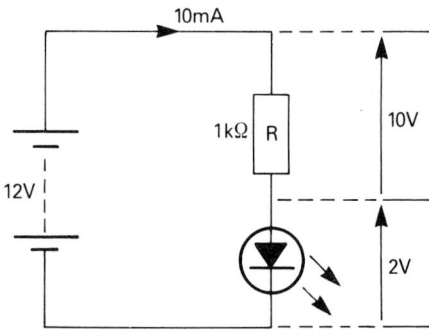

Figure 6.10 Circuit diagram for LED example

about 2 V. If a voltage greater than 2 V is to be used then a resistor must be connected in series with the LED.

To calculate the value of the series resistor we must ask ourselves what we know about LEDs. We know that the diode requires a forward voltage of about 2 V and a current of about 10 mA must flow through the junction to give sufficient light. The value of the series resistor R will, therefore, be given by

$$R = \frac{\text{supply voltage} - 2\,\text{V}}{10\,\text{mA}}\,\Omega$$

Example 1

Calculate the value of the series resistor required when an LED is to be used to show the presence of a 12 V supply.

$$R = \frac{12\,\text{V} - 2\,\text{V}}{10\,\text{mA}}\,\Omega$$

$$R = \frac{10\,\text{V}}{10\,\text{mA}} = 1\,\text{k}\Omega$$

The circuit is, therefore, as shown in Figure 6.10.

LEDs are available in red, yellow and green and when used with a series resistor may replace a filament lamp. They use less current than a filament lamp, are smaller, do not become hot and

last indefinitely. A filament lamp, however, is brighter and emits white light. LEDs are often used as indicator lamps, to indicate the presence of a voltage. They do not, however, indicate the *precise* amount of voltage present at that point.

Another application of the LED is the seven-segment display used as a numerical indicator in calculators, digital watches and measuring instruments. Seven LEDs are arranged as a figure eight so that when various segments are illuminated, the numbers 0 to 9 are displayed as shown in Figure 6.11.

Light-dependent resistor (LDR)

Almost all materials change their resistance with a change in temperature. Light energy falling on a suitable semiconductor material also causes a change in resistance. The semiconductor material of an LDR is encapsulated as shown in Figure 6.12 together with the circuit symbol. The resistance of an LDR in total darkness is about $10\,\text{M}\Omega$, in normal room lighting about $5\,\text{k}\Omega$ and in bright sunlight about $100\,\Omega$. They can carry tens of milli-amperes, an amount which is sufficient to operate a relay. The LDR uses this characteristic to switch on automatically street lighting and security alarms.

Photodiode

The photodiode is a normal junction diode with a transparent window through which light can enter. The circuit symbol and general appearance are

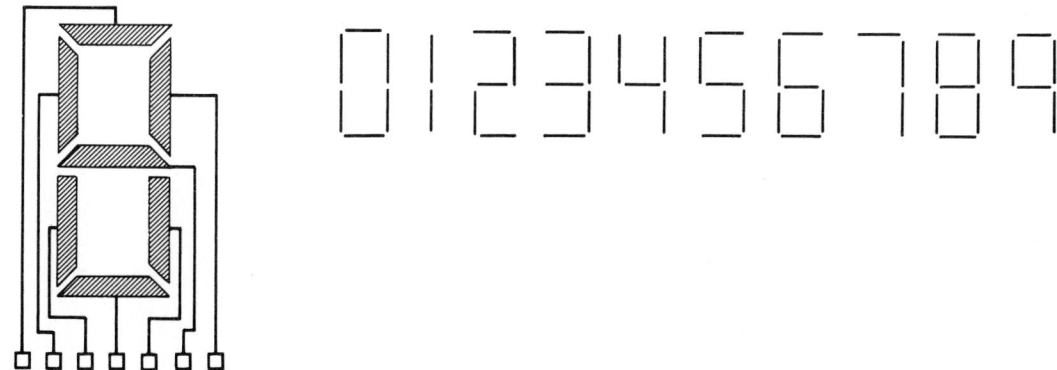

Figure 6.11 LED used in seven-segment display

Symbol

Figure 6.12 Symbol and appearance of a light-dependent resistor

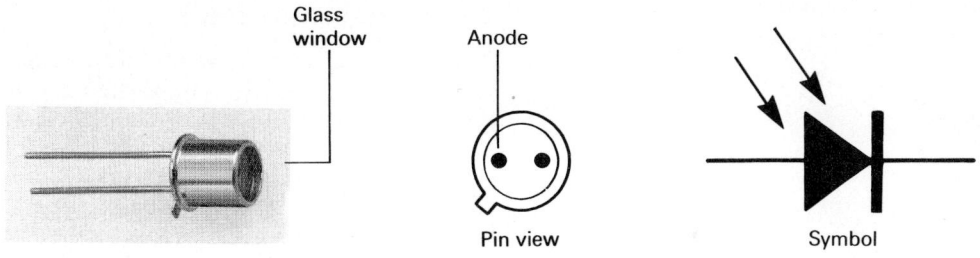

Glass window

Anode

Pin view

Symbol

Figure 6.13 Symbol, pin connections and appearance of a photodiode

shown in Figure 6.13. It is operated in reverse bias mode and the leakage current increases in proportion to the amount of light falling on the junction. This is due to the light energy breaking bonds in the crystal lattice of the semiconductor material to produce holes and electrons.

Photodiodes will only carry micro-amperes of current but can operate much more quickly than LDRs and are used as 'fast' counters when the light intensity is changing rapidly.

Thermistor

The thermistor is a thermal resistor, a semiconductor device whose resistance varies with temperature. The circuit symbol and general appearance are shown in Figure 6.14. They can be supplied in many shapes and are used for the measurement and control of temperature up to their maximum useful temperature limit of about 300°C. They are very sensitive and because the bead of semicon-

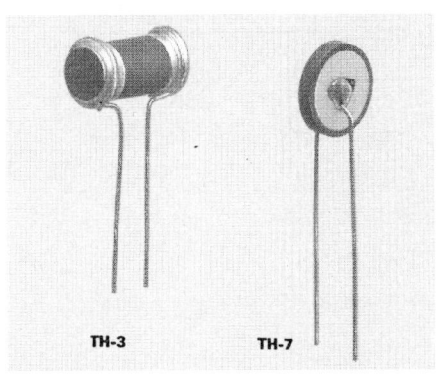
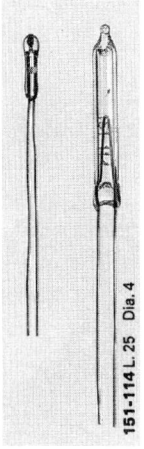
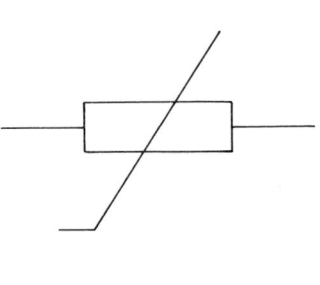

Figure 6.14 Symbol and appearance of a thermistor

ductor material can be made very small, they can measure temperature in the most inaccessible places with very fast response times. Thermistors are embedded in high-voltage underground transmission cables in order to monitor the temperature of the cable. Information about the temperature of a cable allows engineers to load the cables more efficiently. A particular cable can carry a larger load in winter for example, when heat from the cable is being dissipated more efficiently. A thermistor is also used to monitor the water temperature of a motor car.

Transistors

The transistor has become the most important single building block in electronics. It is the modern, miniature, semiconductor equivalent of the thermionic valve and was invented in 1947 by Bardeen, Shockley and Brattain at the Bell Telephone Laboratories in the USA. Transistors are packaged as separate or *discrete* components as shown in Figure 6.15.

There are two basic types of transistor, the *bipolar* or junction transistor and the *Field Effect Transistor* (FET).

The FET has some characteristics which make it a better choice in electronic switches and amplifiers. It uses less power and has a higher resistance and frequency response. It takes up less space than a bipolar transistor and, therefore, more of them can be packed together on a given area of silicon chip. It is, therefore, the FET which

is used when many transistors are integrated on to a small area of silicon chip as in the *integrated circuit* (IC) discussed later.

When packaged as a discrete component the FET looks much the same as the bipolar transistor. The circuit symbol and connections are given in the Appendix. However, it is the bipolar transistor which is much more widely used in electronic circuits as a discrete component.

The bipolar transistor

The bipolar transistor consists of three pieces of semiconductor material sandwiched together as shown in Figure 6.16. The structure of this transistor makes it a three-terminal device having a base, collector and emitter terminal. By varying the current flowing into the base connection a much larger current flowing between collector and emitter can be controlled. Apart from the supply connections, both npn and pnp types are essentially the same but the npn type is more common.

A transistor is generally considered a current-operated device. There are two possible current paths through the transistor circuit, shown in Figure 6.17, the base emitter path when the switch is closed and the collector emitter path. Initially, the positive battery supply is connected to the n-type material of the collector, the junction is reverse biased and, therefore, no current will flow. Closing the switch will forward bias the base-emitter junction and current flowing through this junction causes current to flow across the collector-emitter junction and the signal lamp will light.

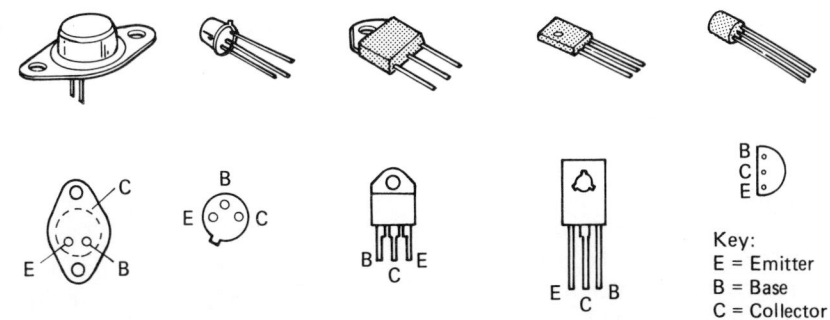

Transistor families

Key:
E = Emitter
B = Base
C = Collector

Figure 6.15 The appearance and pin connections of the transistor family

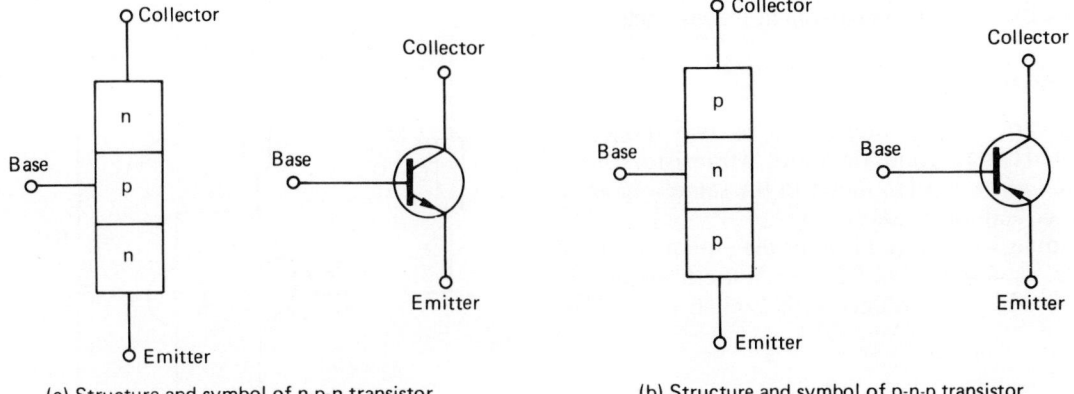

(a) Structure and symbol of n-p-n transistor

(b) Structure and symbol of p-n-p transistor

Figure 6.16 Structure and symbol of npn and pnp transistors

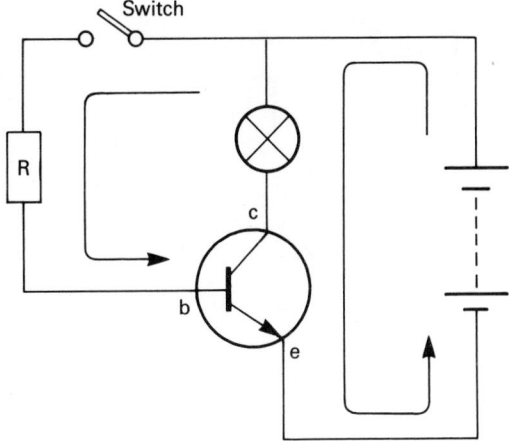

Figure 6.17 Operation of the transistor

A small base current can cause a much larger collector current to flow. This is called the *current gain* of the transistor and is typically about 100. When I say a much larger collector current, I mean a large current in electronic terms, up to about half an ampere.

We can, therefore, regard the transistor as operating in two ways, as a switch because the base current turns on and controls the collector current, and secondly as a current amplifier because the collector current is greater than the base current.

We could also consider the transistor to be operating in a similar way to a relay. However, the transistor has many advantages over electrically operated switches such as relays. They are very small, reliable, have no moving parts and, in particular, they can switch millions of times a second without arcing occurring at the contacts.

Transistor testing

A transistor can be thought of as two diodes connected together and, therefore, a transistor can be tested using an ohm meter in the same way as was described for the diode.

Assuming that the red lead of the ohm meter is positive, as described in Chapter 5, the transistor can be tested in accordance with Table 6.1.

When many transistors are to be tested, a simple test circuit can be assembled as shown in Figure 6.18.

With the circuit connected, as shown in Figure 6.18, a 'good' transistor will give readings on the voltmeter of 6 V with the switch open and about 0.5 V when the switch is made. The voltmeter used for the test should have a high internal resistance, about ten times greater than 4.7 kΩ, this is usually indicated on the back of a multi-range meter or in the manufacturers' information supplied with a new meter.

Table 6.1   Transistor testing using an Ohm Meter

A 'good' NPN transistor will give the following readings

Red to base and black to collector = low resistance
Red to base and black to emitter = low resistance

Reversed connections on the above terminals will result in a high resistance reading as will connections of either polarity between the collector and emitter terminals.

A 'good' PNP transistor will give the following readings

Black to base and Red to collector = low resistance
Black to base and Red to emitter = low resistance

Reversed connections on the above terminals will result in a high resistance reading as will connections of either polarity between the collector and emitter terminals.

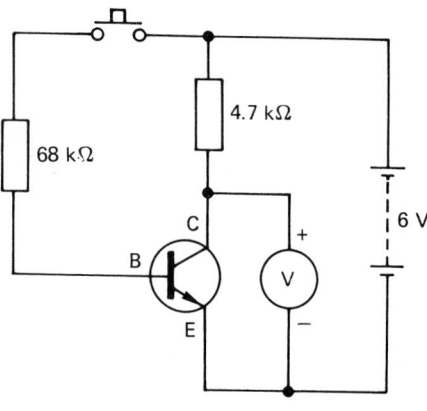

n-p-n transistor test

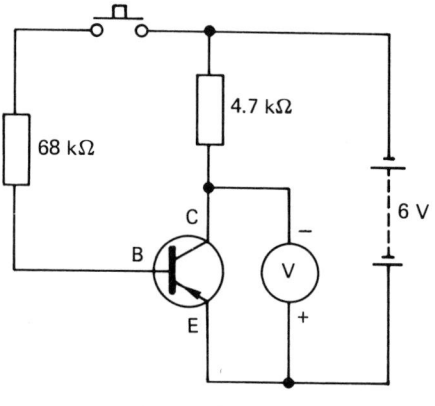

p-n-p transistor test

Figure 6.18 Transistor test circuits

Integrated circuits

Integrated circuits were first developed in the 1960s. They are densely populated miniature electronic circuits made up of hundreds and sometimes thousands of microscopically small transistors, resistors, diodes and capacitors, all connected together on a single chip of silicon no bigger than a baby's little finger nail. When assembled in a single package, as shown in Figure 6.19, we call the device an integrated circuit.

There are two broad groups of integrated circuit, digital ICs and linear ICs. Digital ICs contain simple switching-type circuits used for logic control and calculators, discussed in Chapter 8. Linear ICs incorporate amplifier-type circuits which can respond to audio and radio frequency signals. The most versatile linear IC is the operational amplifier which has applications in electronics, instrumentation and control, and its application to strain gauges is discussed in Chapter 9.

The integrated circuit is the electronic revolution. ICs are more reliable, cheaper, smaller and electronically superior to the same circuit made from discrete or separate transistors. One IC behaves differently to another because of the arrangement of the transistors within the IC.

Manufacturers' data sheets describe the characteristics of the different ICs which have a reference number stamped on the top. The Appendix gives the characteristics of some of the more common ICs.

When building circuits, it is necessary to be able to identify the IC pin connection by number. The number one pin of any IC is indicated by a dot pressed into the encapsulation or, the number one pin is the pin to the left of the cutout as shown in Figure 6.20. Since the packaging of ICs has two rows of pins they are called DIL (Dual In Line) packaged integrated circuits and their appearance is shown in Figure 6.21.

Integrated circuits are sometimes connected into DIL sockets and at other times are soldered directly into the circuit. The testing of ICs is beyond the scope of a practising electrician and when they are suspected of being faulty an identical or equivalent replacement should be connected

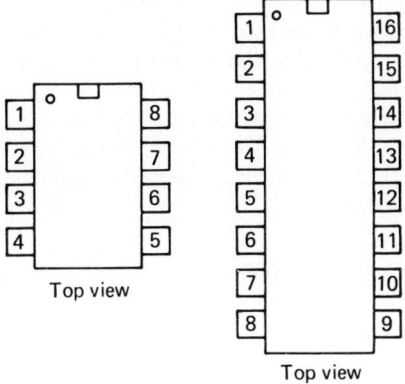

Figure 6.20 IC pin identification

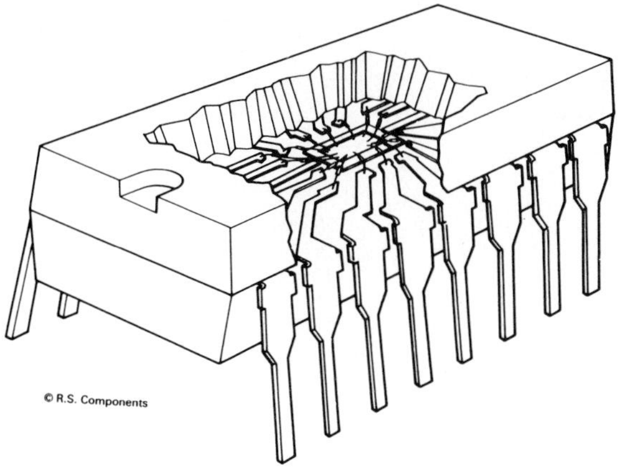

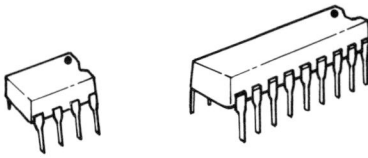

Figure 6.21 DIL packaged integrated circuits

Figure 6.19 Exploded view of an integrated circuit

into the circuit, ensuring that it is inserted the correct way round, which is indicated by the position of pin number one as described earlier.

The Thyristor or Silicon-controlled Rectifier (SCR)

The thyristor was previously known as a *silicon controlled rectifier* since it is a rectifier which controls the power to a load. It consists of four pieces of semiconductor material sandwiched together and connected to three terminals as shown in Figure 6.22.

The word thyristor is derived from the Greek word *thyra* meaning door, because the thyristor behaves like a door. It can be open or shut, allowing or preventing current flow through the device. The door is opened, or we say the thyristor is triggered, to a conducting state by applying a pulse voltage to the gate connection. Once the thyristor is in the conducting state, the gate loses all control over the device. The only way to bring the thyristor back to a non-conducting state is to reduce the voltage across the anode and cathode to zero or apply reverse voltage across the anode and cathode.

We can understand the operation of a thyristor by considering the circuit shown in Figure 6.23. This circuit can also be used to test suspected faulty components.

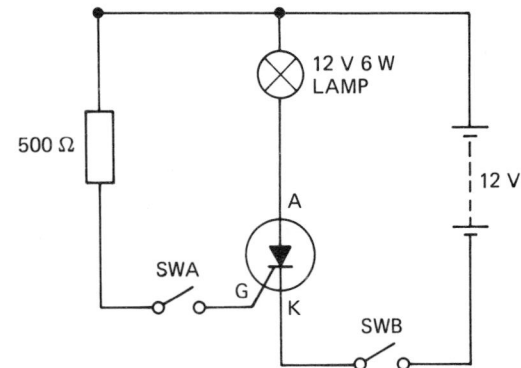

Figure 6.23 Thyristor test circuit

Thyristor operation and testing

When SW.B only is closed the lamp will not light, but when SW.A is also closed, the lamp lights to full brilliance. The lamp will remain illuminated even when SW.A is opened. This shows that the thyristor is operating correctly. Once a voltage has been applied to the gate the thyristor becomes forward conducting, like a diode, and the gate loses control.

Thyristor in practice

The thyristor has no moving parts and operates without arcing. It can operate at extremely high speeds and the currents used to operate the gate

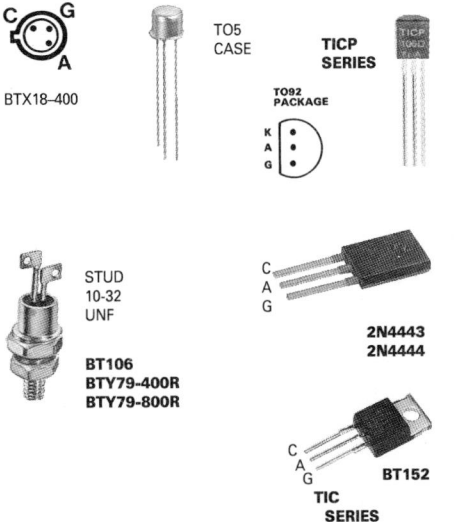

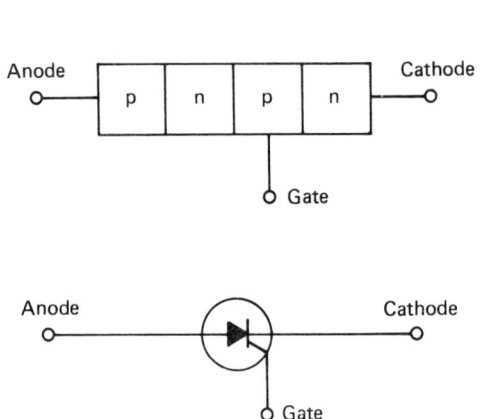

Figure 6.22 Structure, symbol and appearance of a thyristor

are very small. The most common application for the thyristor is to control the power supply to a load, for example, lighting dimmers and motor speed control. A number of circuits are considered in Chapter 7.

The power available to an a.c. load can be controlled by allowing current to be supplied to the load during only a part of each cycle. This can be achieved by supplying a gate pulse automatically at a chosen point in each cycle as shown by Figure 6.24. Power is reduced by triggering the gate later in the cycle.

The thyristor is only a half-wave device (like a diode) allowing control of only half the available power in an a.c. circuit. This is very uneconomical and a further development of this device has been the triac which is considered next.

Thyristor testing using an ohm meter

A thyristor may also be tested using an ohm meter as described in Table 6.2, assuming that the red lead of the ohm meter is positive as described in Chapter 5.

The triac

The triac was developed following the practical problems experienced in connecting two thyristors in parallel, to obtain full wave control, and in providing two separate gate pulses to trigger the two devices.

The triac is a single device containing a back-to-back, two-directional thyristor which is triggered on both halves of each cycle of the a.c. supply by the same gate signal. The power available to the load can, therefore, be varied between zero and full load.

The symbol and general appearance are shown in Figure 6.25. Power to the load is reduced by triggering the gate later in the cycle as shown by the waveforms of Figure 6.26.

The triac is a three-terminal device, just like the thyristor, but the terms anode and cathode have no meaning for a triac. Instead, they are called main terminal one (MT1) and main terminal two (MT2). The device is triggered by applying a small pulse to the gate (G). A gate current of 50 mA is sufficient to trigger a triac switching up to 100 A. They are used for many commercial applications

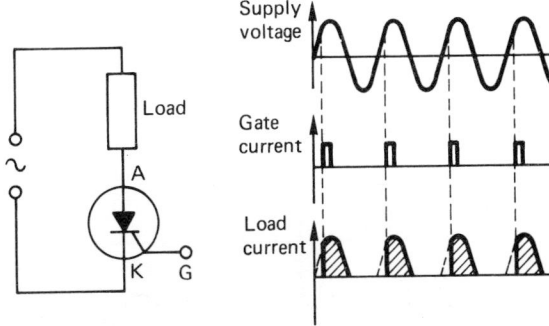

Figure 6.24 Waveforms to show the control effect of a thyristor

Table 6.2 Thyristor Testing using an Ohm Meter

A 'good' thryistor will give the following readings

Black to cathode and red on gate = low resistance
Red to cathode and black on gate = a higher resistance value

The value of the second reading will depend upon the thyristor, and may vary from only slightly greater to very much greater.

From cathode to anode with either polarity connected will result in a very high resistance reading.

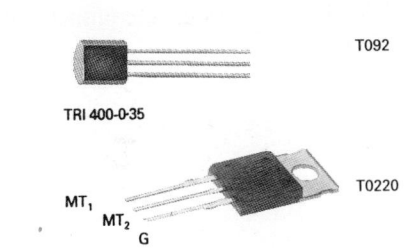

Figure 6.25 Appearance of a triac

where control of a.c. power is required, for example, motor speed control and lamp dimming.

The diac

The diac is a two-terminal device containing a two-directional zener diode. It is used mainly as a

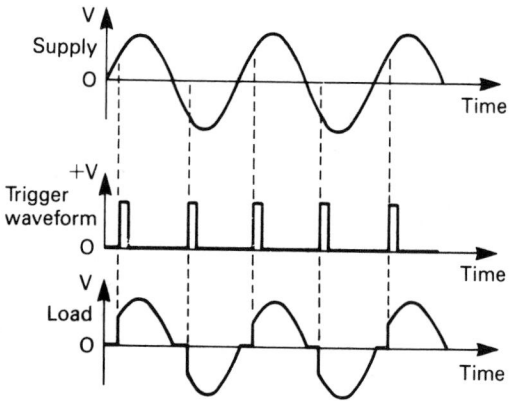

Figure 6.26 Waveforms to show the control effect of a triac

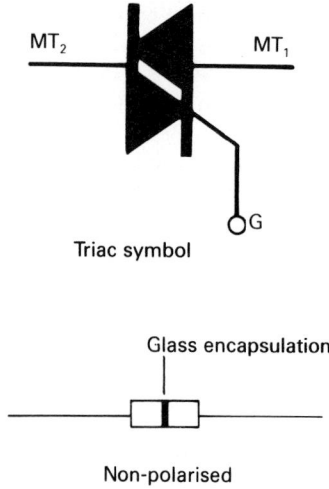

Triac symbol

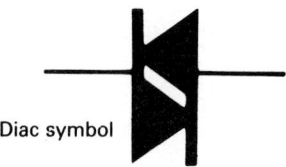

Glass encapsulation

Non-polarised

Diac symbol

trigger device for the thyristor and triac. The symbol is shown in Figure 6.27.

The device turns on when some predetermined voltage level is reached, say 30 V, and, therefore, it can be used to trigger the gate of a triac or thyristor each time the input waveform reaches this predetermined value. Since the device contains back-to-back zener dodes it triggers on both the positive and negative half cycles.

The applications of these semiconductor devices in various electronic circuits are considered in Chapter 7.

Manufacturers' data sheet information and practical information on these semiconductor devices is given in the Appendix.

Figure 6.27 The symbol and appearance of a diac used in triac firing circuits

CHAPTER 7

Electronic circuits

Voltage divider

In Chapter 3 we considered the distribution of voltage across resistors connected in series. We found that the supply voltage was divided between the series resistors in proportion to the size of the resistor. If two identical resistors were connected in series across a 12 V supply as shown in Figure 7.1 (a) both common sense and a simple calculation would confirm that 6 V would be measured across the output. In the circuit shown in Figure 7.1(b) the 1 kΩ and 2 kΩ resistors divide the input voltage into three equal parts. One part, four volts, will appear across the 1 kΩ resistor and two parts, eight volts, will appear across the 2 kΩ resistor. In Figure 7.1(c) the situation is reversed and, therefore, the voltmeter will read 4 V. The division of the voltage is proportional to the ratio of the two resistors and, therefore, we call this simple circuit a voltage divider or potential divid-

er. The values of the resistors R_1 and R_2 determine the output voltage as follows:

$$V_{OUT} = V_{IN} \times \frac{R_2}{R_1 + R_2} \text{ Volts}$$

For the circuit shown in Figure 7.1(b)

$$V_{OUT} = 12\text{ V} \times \frac{2\text{ k}\Omega}{1\text{ k}\Omega + 2\text{ k}\Omega} = 8\text{ V}$$

For the circuit shown in Figure 7.1(c)

$$V_{OUT} = 12\text{ V} \times \frac{1\text{ k}\Omega}{2\text{ k}\Omega + 1\text{ k}\Omega} = 4\text{ V}$$

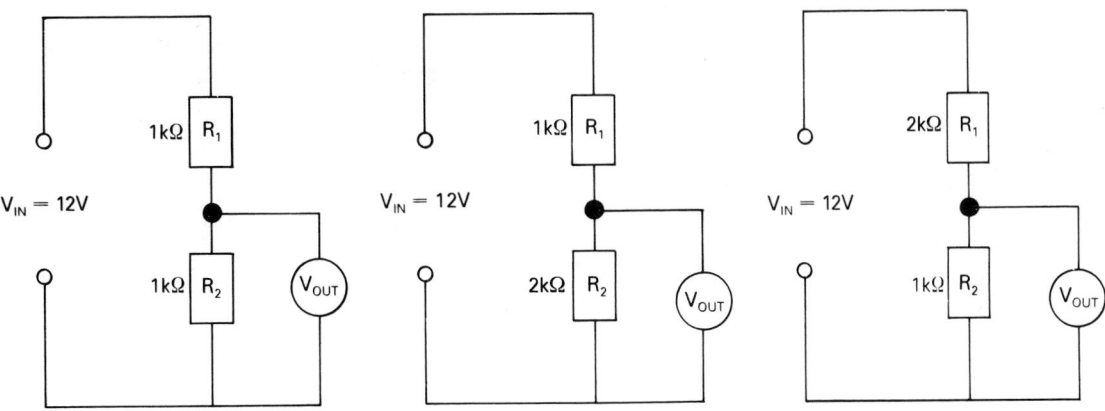

Figure 7.1 Voltage divider circuit

Example 1

For the circuit shown in Figure 7.2 calculate the output voltage

$$V_{OUT} = 6\,V \times \frac{2.2\,k\Omega}{10\,k\Omega + 2.2\,k\Omega} = 1.08\,V$$

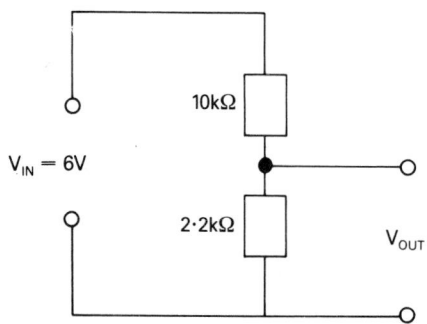

Figure 7.2 Voltage divider circuit for Example 1

Example 2

For the circuit shown in Figure 7.3(a) calculate the output voltage. We must first calculate the equivalent resistance of the parallel branch

$$\frac{1}{R_T} = \frac{1}{R_1} + \frac{1}{R_2}$$

$$\frac{1}{R_T} = \frac{1}{10\,k\Omega} + \frac{1}{10\,k\Omega} = \frac{1+1}{10\,k\Omega} = \frac{2}{10\,k\Omega}$$

$$R_T = \frac{10\,k\Omega}{2} = 5\,k\Omega$$

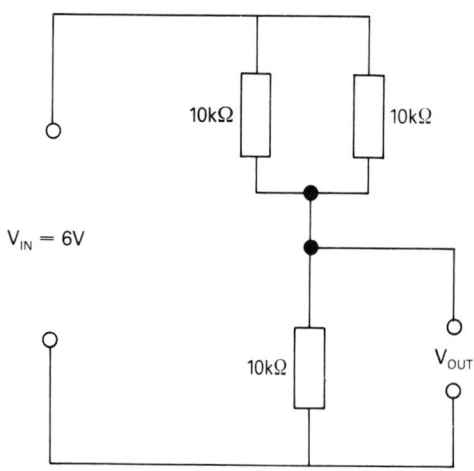

Figure 7.3 Voltage divider circuit for Example 2

The circuit may now be considered as shown in Figure 7.3(b)

$$V_{OUT} = 6\,V \times \frac{10\,k\Omega}{5\,k\Omega + 10\,k\Omega} = 4\,V$$

Voltage dividers are used in electronic circuits to produce a reference voltage which is suitable for operating transistors and integrated circuits. The volume control in a radio or the brightness control of a CRO requires a continuously variable voltage divider and this can be achieved by connecting a variable resistor or potentiometer as shown in Figure 7.4. With the wiper arm making a connec-

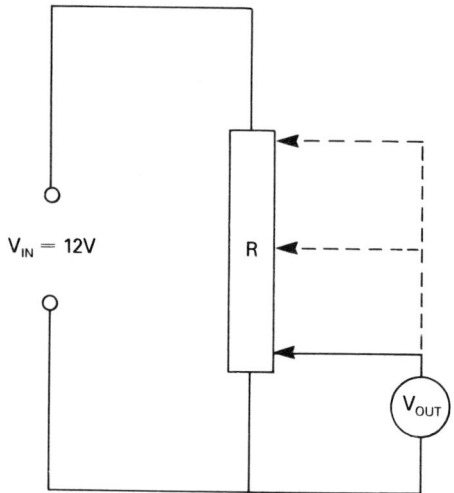

Figure 7.4 Constantly variable voltage divider circuit

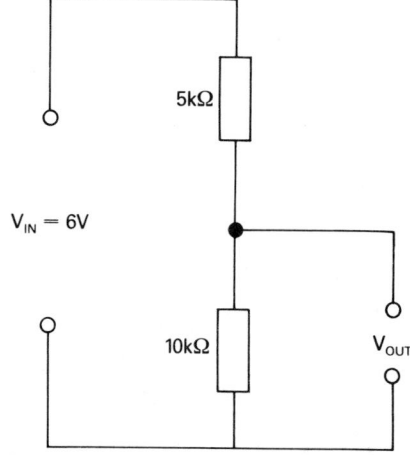

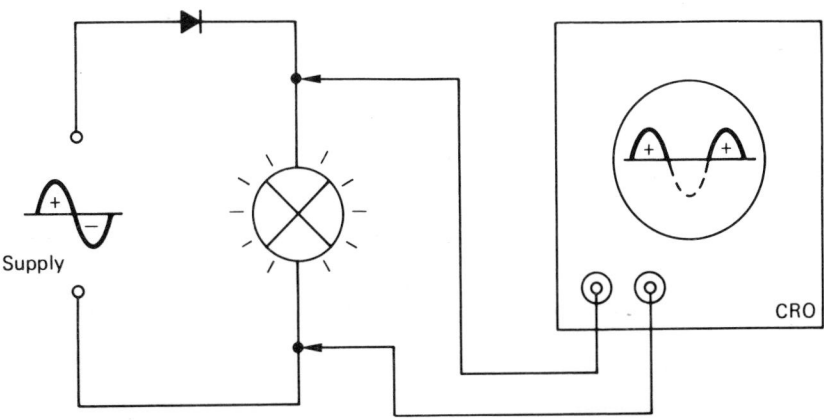

Figure 7.5 Half-wave rectification

tion at the bottom of the resistor, the output would be zero volt. When connection is made at the centre, the voltage would be six volts, and at the top of the resistor the voltage would be 12 volts. The voltage is continuously variable between 0 V and 12 V simply by moving the wiper arm of a suitable variable resistor such as those shown in Figure 1.3.

When a load is connected to a voltage divider it 'loads' the circuit, causing the output voltage to fall below the calculated value. To avoid this, the resistance of the load should be at least ten times greater than the value of the resistor across which it is connected. For example, the load connected across the voltage divider shown in Figure 7.1(b) must be greater than $20 \, k\Omega$ and across 7.1(c) greater than $10 \, k\Omega$. This problem of loading the circuit also occurs when taking voltage readings as we discussed in Chapter 5.

Rectification of a.c.

When a d.c. supply is required, batteries or a rectified a.c. supply can be provided. Batteries have the advantage of portability but a battery supply is more expensive than using the a.c. mains supply suitably rectified. Rectification is the conversion of an alternating current supply into a uni-directional or direct current supply. This is one of the many applications for a diode which will conduct in one direction only, that is when the anode is positive with respect to the cathode as we discussed in Chapter 6.

Half-wave rectification

The circuit is connected as shown in Figure 7.5. During the first half cycle the anode is positive with respect to the cathode and, therefore, the diode will conduct. When the supply goes negative during the second half cycle, the anode is negative with respect to the cathode and, therefore, the diode will not allow current to flow. Only the positive half of the waveform will be available at the load and the lamp will light at reduced brightness.

Full-wave rectification

Figure 7.6 shows an improved rectifier circuit which makes use of the whole a.c. waveform and is, therefore, known as a full-wave rectifier. When the four diodes are assembled in this diamond-shaped configuration, the circuit is also known as a *bridge rectifier*. During the first half cycle diodes D_1 and D_3 conduct and diodes D_2 and D_4 conduct during the second half cycle. The lamp will light to full brightness.

Full-wave and half-wave rectification can be displayed on the screen of a CRO and will appear as shown in Figures 7.5 and 7.6.

Smoothing

The circuits of Figures 7.5 and 7.6 convert an alternating waveform into a waveform which never

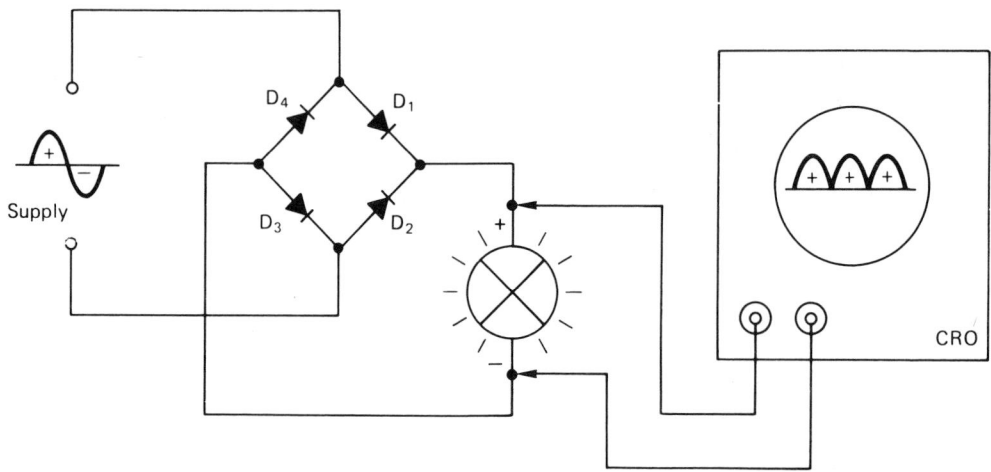

Figure 7.6 Full-wave rectification using a bridge circuit

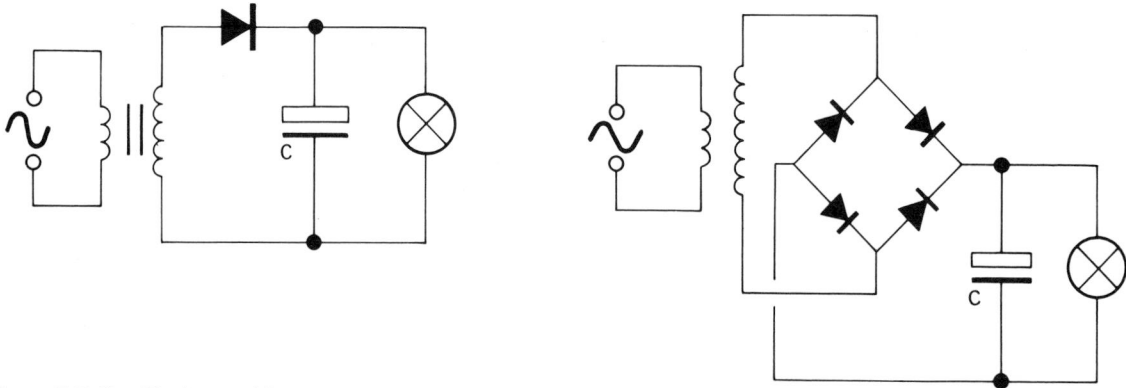

Figure 7.7 Rectified a.c. with smoothing capacitor connected

goes negative, but they cannot be called continuous d.c. because they contain a large alternating component. Such a waveform is too bumpy to be used to supply electronic equipment but may be used for battery charging. To be useful in electronic circuits the output must be smoothed. The simplest way to smooth an output is to connect a large-value capacitor across the output terminals as shown in Figure 7.7.

When the output from the rectifier is increasing, as shown by the dotted lines of Figure 7.8, the capacitor charges up. During the second quarter of the cycle, when the output from the rectifier is falling to zero, the capacitor discharges into the load. The output voltage falls until the output from the rectifier once again charges the capacitor. The capacitor connected to the full wave rectifier circuit is charged up twice as often as the capacitor connected to the half-wave circuit and, therefore, the output ripple on the full-wave circuit is smaller, giving better smoothing. Increasing the current drawn from the supply increases the size of the ripple. Increasing the size of the capacitor reduces the amount of ripple.

Low pass filter

The ripple voltage of the rectified and smoothed circuit shown in Figure 7.7 can be further reduced by adding a low pass filter, as shown in Figure 7.9. A low pass filter allows low frequencies to pass while blocking higher frequencies. D.c. has a frequency of zero Hz while the ripple voltage of a full-wave rectifier has a frequency of 100 Hz.

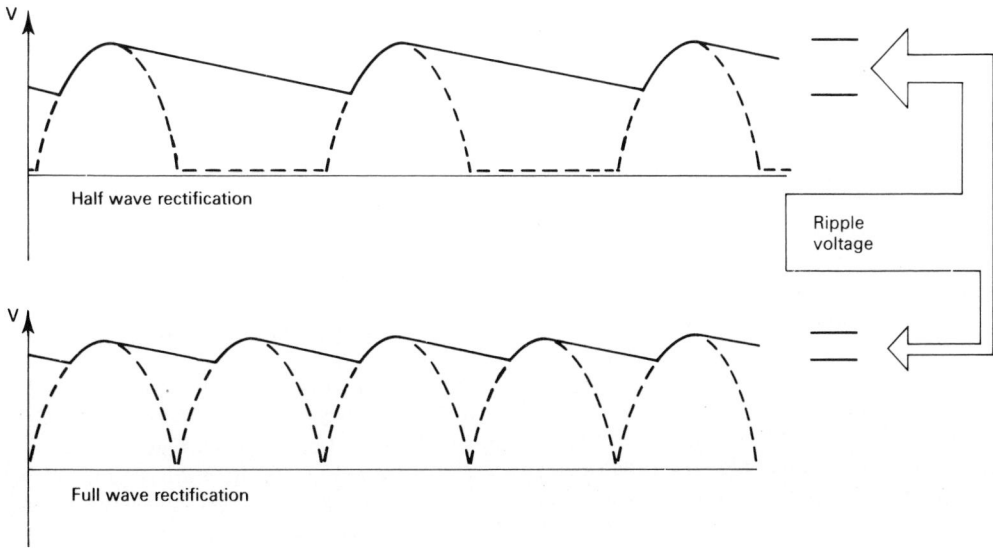

Figure 7.8 Output waveforms with smoothing showing reduced ripple with fullwave

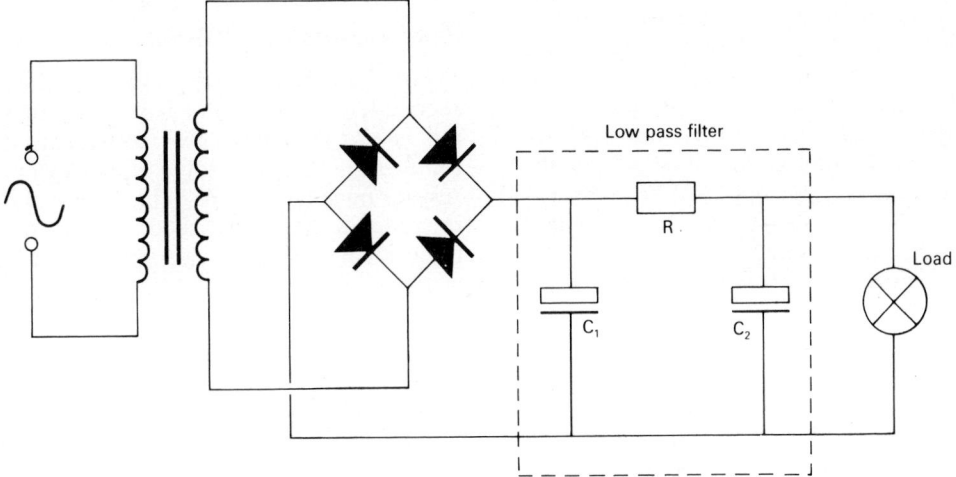

Figure 7.9 Rectified a.c. with low pass filter connected

Connecting the low pass filter will allow the d.c. to pass while blocking the ripple voltage, resulting in a smoother output voltage.

The low pass filter shown in Figure 7.9 does, however, increase the output resistance which encourages the voltage to fall as the load current increases. This can be reduced if the resistor is replaced by a choke, which has a high impedance to the ripple voltage but a low resistance which reduces the output ripple without increasing the output resistance.

Diode and capacitor ratings

The selection of diodes and capacitors for a rectified a.c. supply needs some careful consideration. It is important to choose a diode which will not be damaged by either the current flowing through it or the voltage developed across it. The capacitor discharges very little during the half cycle that the transformer changes from its maximum positive to maximum negative value and, therefore, the voltage across the diode can be equal to almost twice the maximum value of the a.c. supply. The diode must withstand this maximum reverse voltage without damage, and, therefore, a diode must be chosen which has a maximum reverse voltage rating of about four times the rms voltage of the transformer secondary voltage. This will then take account of the maximum reverse voltage plus any mainsborne interference which may be present.

The capacitor should have a voltage rating, at least equal to the peak value of the transformer secondary voltage, so that the bumpy output from the diodes will not break down the capacitor insulation. Also, over much of the a.c. cycle, the load current is supplied by the capacitor, which must be capable of supplying the load without overheating. The 'ripple current rating' of a capacitor is usually quoted in the manufacturers' specification, which for an electrolytic capacitor may range from 200 mA upwards.

When the load current of the simple circuit shown in Figure 7.7 increases, the terminal voltage reduces, mainly because of the resistance of the transformer secondary winding. This reduction in the terminal voltage is called the load regulation or just 'regulation' of the circuit.

Regulation

Regulation is the percentage change in the output voltage when the load current is increased from zero to full load. A power supply whose output voltage falls rapidly as the load current is increased would be said to have poor regulation. To overcome poor load regulation a number of regulating or stabilising circuits have been developed.

Stabilised power supplies

The power supplies required for electronic circuits must be ripple free, stabilised and have good regulation, that is the voltage must not change in value over the whole load range. A number of stabilising circuits are available which, when connected across the output of the circuit shown in Figure 7.7, give a constant or stabilised voltage output. These circuits use the characteristics of the zener diode which was discussed in Chapter 6.

Zener diode stabiliser

Figure 7.10 shows an a.c. supply which has been rectified, smoothed and stabilised. In designing the circuit we must give consideration to the following points. The value of the current-limiting resistor must be chosen so that the power rating of the zener diode is not exceeded when operating. In the case of a BZY88-type zener diode, the max-

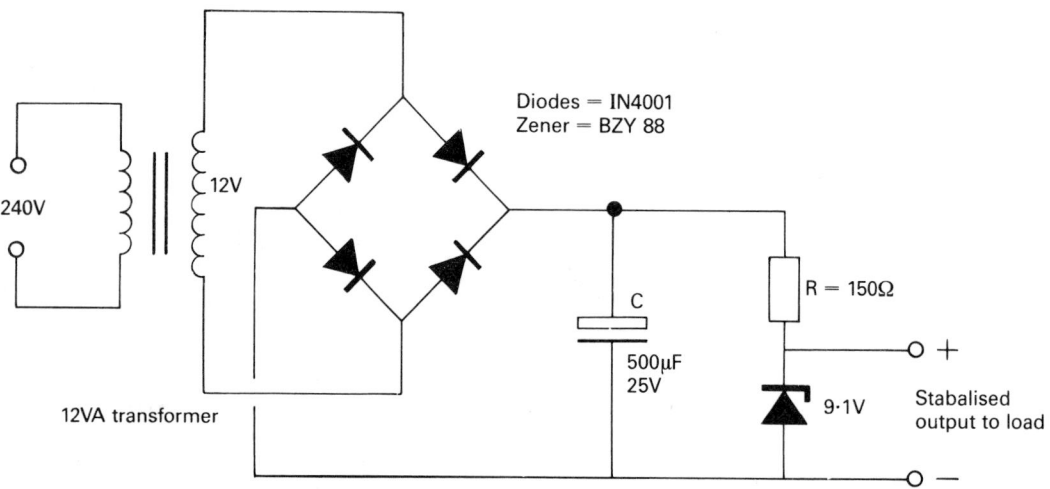

Figure 7.10 Stabilised d.c. supply

imum power rating is 500 mW and, therefore, the maximum current which can flow through a 9.1 V zener diode will be given by

$$I_{max} = \frac{\text{Power W}}{\text{Volts V}} = \frac{500 \times 10^{-3}\ \text{W}}{9.1\ \text{V}} = 55\ \text{mA}.$$

When there is no load connected the output current from the smoothing circuit will flow through the zener diode, and this must be limited to no more than 55 mA by the series resistor, to prevent the diode overheating. However, because the power dissipation of the zener diode is determined for free air, we will base further calculations upon a maximum current of only 40 mA to avoid any possibility of the diode overheating. The output voltage from the smoothing capacitor will be greater than the rms voltage of the transformer secondary because the capacitor will charge up to the maximum value of the secondary voltage. The voltage across the capacitor of Figure 7.10 will, therefore, be approximately 15 V after deducting the volt drop across the two diodes from the maximum value. This voltage will become the input voltage to the stabilising circuit and the voltage levels are shown in Figure 7.11.

If we assume that 40 mA flows through the series resistor and diode, the resistor must have a value of

$$R = \frac{V}{I} = \frac{6\ \text{V}}{40 \times 10^{-3}\ \text{A}} = 150\ \Omega$$

The smoothing capacitor must be capable of supplying the ripple volts and have a voltage rating of at least the peak transformer voltage.

The value of all the circuit components are as shown on the diagram given in Figure 7.10.

When the load current is a maximum, sufficient current must flow through the zener diode to turn it on to the breakdown part of the characteristic shown in Figure 6.7. This will be at least 5 mA for a small diode and is usually achieved by making the input to the stabilising circuit several volts greater than the output.

The circuit shown in Figure 7.10 will, therefore, provide a stabilised voltage output of approximately 9 V and be capable of delivering a maximum load current of 35 mA.

The output voltage of this simple stabilised circuit may vary slightly because the slope of the

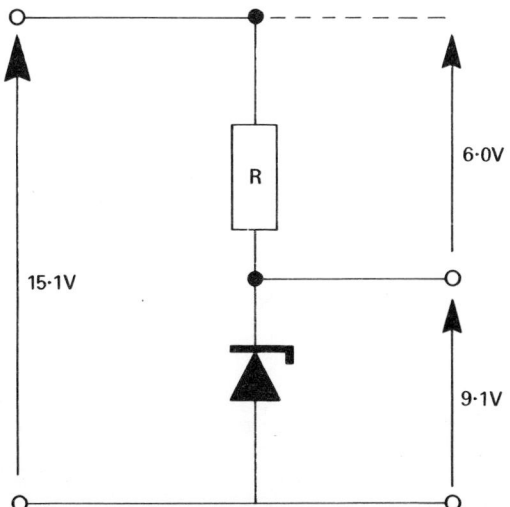

Figure 7.11 Voltage levels for the stabilising circuit given in Fig.7.10

zener diode characteristic is not perfectly linear. To give a constant output, the current through the zener diode must remain constant. Two ways of achieving this are to use a second diode as a stabiliser or a transistor as a constant current source.

Figure 7.12 shows a slightly improved but more expensive circuit which makes use of a second zener diode to compensate for regulation and, therefore, provide a more stable output.

The circuit shown will, however, only supply a load of about 15 mA and a much more elegant solution is to use a transistor as a constant current source. In this case the output current is multiplied by the current gain of the transistor without increasing the demand on the zener diode. In the case of Figure 7.13 the output will be about one ampere. The smoothing capacitor must be big enough to prevent the ripple voltage dropping below 9.1 V while supplying the load current and, therefore, the parallel capacitors have been used in this circuit to supply the relatively large load. (In electronic circuits one ampere represents a lot of current!)

Fixed voltage series regulators

For really heavy current applications (2 A to 5 A), a voltage regulator connected in series with the

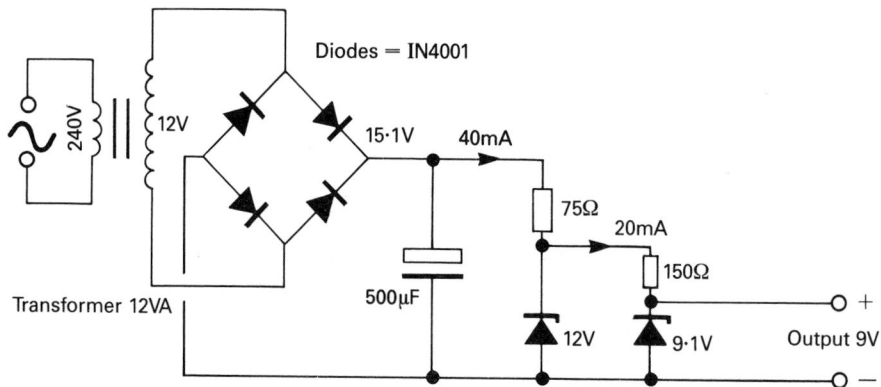

Figure 7.12 Stabilised d.c. supply with compensation for regulation

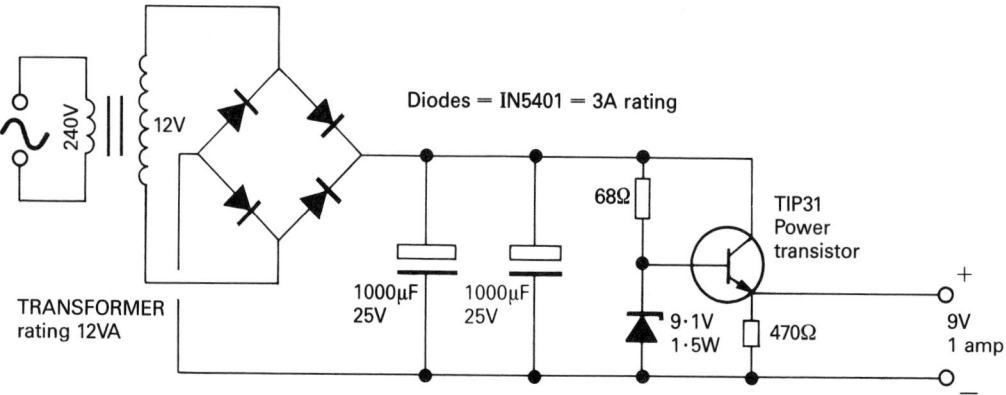

Figure 7.13 Stabilised d.c. power supply

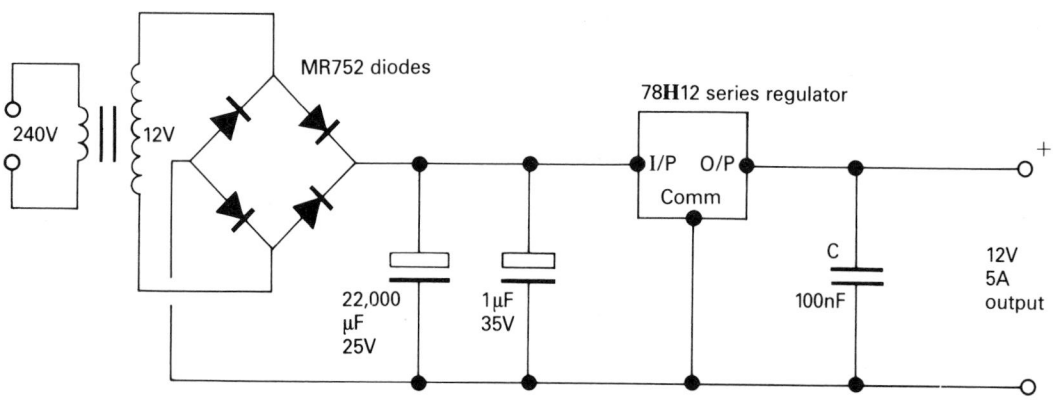

Figure 7.14 Stabilised d.c. power supply using a fixed voltage series regulator

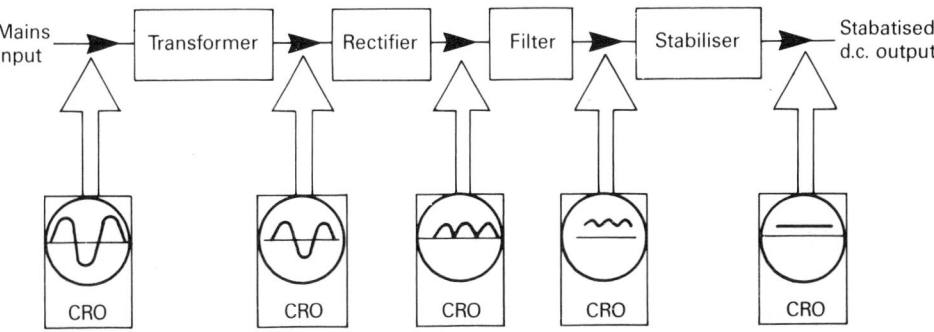

Figure 7.15 A system or block diagram for a d.c. power supply showing the wave shape at each stage

load, is the most elegant and least expensive solution. This is a three-terminal integrated circuit which incorporates zener diodes, transistors, resistors and capacitors in one package. The internal circuitry provides automatic thermal overload protection and to achieve maximum performance the internal power dissipation must be kept below 50 W by mounting the voltage regulator on a heat sink. Figure 7.14 shows the basic circuit diagram for a fixed voltage regulator with an output of 12 V and capable of supplying output currents up to five ampere. Other output voltages and ratings are available as three terminal packages of integrated circuits and these are tabulated in Appendix I.

The system or block diagram for a d.c. power supply is shown in Figure 7.15. You should by now be reasonably familiar with the contents of each section. The transformer converts the a.c. mains supply to a safer, lower value a.c. supply. The rectifier converts the a.c. supply into a bumpy d.c. supply. The filter, usually a high-value electrolytic capacitor, smooths out the bumpy d.c. supply to no more than a ripple. The stabiliser, a circuit incorporating a zener diode, fixes and maintains the voltage level at a precise value. The shape of the waveform as it would appear on an oscilloscope at each stage is also shown in Figure 7.15.

Power control using the thyristor

The thyristor is a power rectifier which is triggered into its forward conducting state by applying a small positive pulse or continuous d.c. voltage to the 'gate' connection. It is also called a silicon *controlled* rectifier (SCR) since the gate pulse *controls* the switching of the device which is made from silicon.

Once switched into the forward conducting state by the gate pulse, the device cannot be switched off until the applied voltage falls to zero. Thyristors, therefore, lend themselves to the control of power in a.c. circuits because the voltage waveform of the a.c. mains passes through zero twice on each cycle, which allows the device to be switched off. Two methods are generally employed to control a thyristor in an a.c. circuit, burst trigger control or phase control.

Burst trigger control

Burst triggering, synchronous triggering or zero voltage triggering are different names for the same method of thyristor control which is directly comparable with traditional methods of control in which power is switched on and off for various intervals of time as shown in Figure 7.16. In this case the thyristor is triggered only on the first, third and fourth cycles, but more power could be delivered to the load by triggering every cycle, or less power by triggering fewer cycles. Mechanical or thermal inertia is used to smooth out the effects of this bumpy waveform on the load. Switching the thyristor on as the mains waveform passes through zero has the advantage of reducing the effect of mainsborne interference which occurs if the thyristor is switched on part way through a cycle but a much smoother and more desirable method of controlling the mains waveform is provided by phase control.

Phase control

Phase control is a technique used for varying the effective power to a load by rapidly switching the

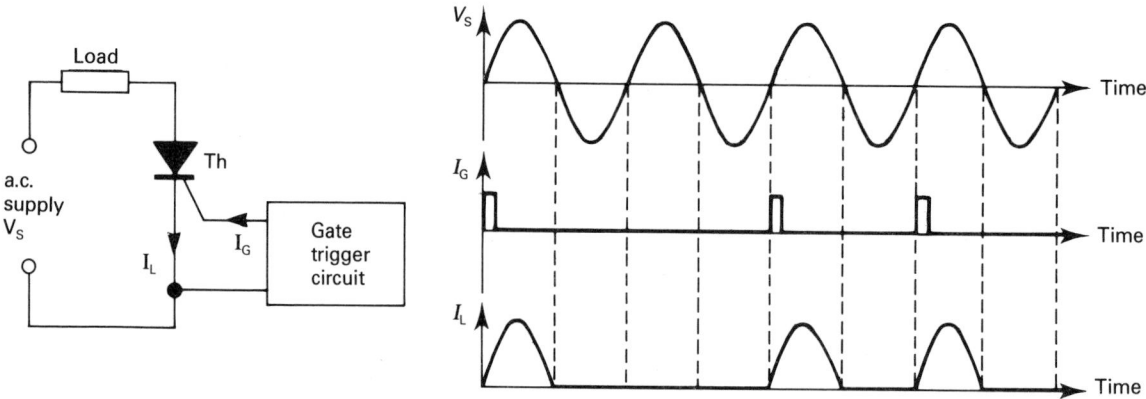

Figure 7.16 Burst trigger control of a thyristor

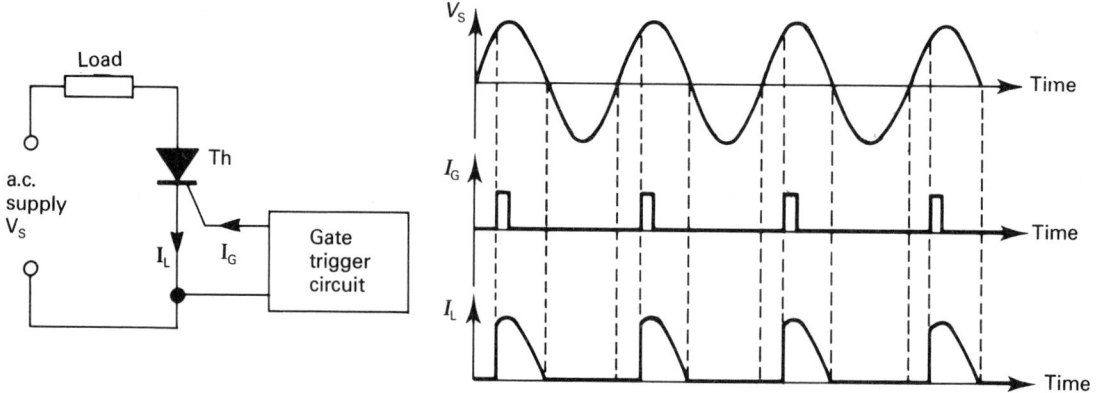

Figure 7.17 Phase control of a thyristor

a.c. supply connected to the load for a controlled and adjustable fraction of every cycle. Triggering the thyristor early in the cycle will deliver more power to the load. Triggering the thyristor later in the cycle will reduce the power available. The angle at which the thyristor is turned on is called the trigger angle α (Greek letter alpha) and the angle through which the thyristor conducts is called the conduction angle θ (Greek letter theta). Figure 7.17 shows a triggering angle of about 45°.

This method of control is different to burst triggering because periods of conduction occur during *every* positive half cycle of the mains voltage. The supply to the load is less bumpy and this method of control is, therefore, ideally suited for applications where a smooth supply of controllable power is required. Phase control is essential if filament lamps are to be dimmed electronically, otherwise lamp flicker is troublesome.

Figure 7.18 shows a simple thyristor-controlled circuit which could be used for motor speed control or as a lamp dimmer.

The trigger angle and, therefore, the power available to the load is adjusted by varying the resistor R_2. Resistors R_1 and R_2 act as a voltage divider for the circuit. The diode D_2 ensures that only a positive voltage is applied to the gate of the thyristor. The diode D_1 prevents the negative half cycle of the mains waveform being presented to the gate of the thyristor.

Adjusting the variable resistor R_2 will vary the voltage at the gate of the thyristor. This will change the trigger angle and, therefore, vary the power available to the load giving, in the case of Figure 7.18, speed control of the electric motor.

The switching of motor and discharge lighting circuits causes problems for electronic circuit designers because of the inductive nature of these

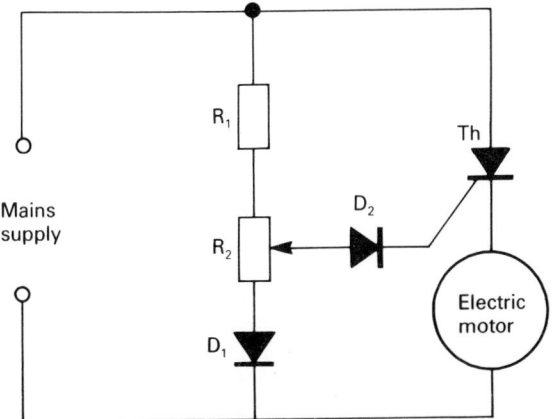

Figure 7.18 A simple thyristor-controlled circuit

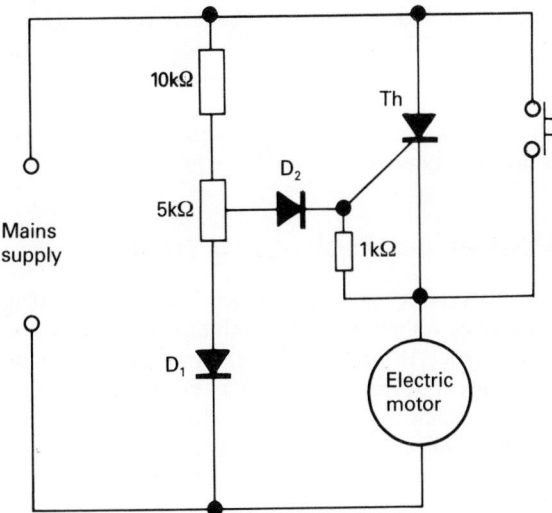

Figure 7.19 Thyristor speed control of an electric motor

loads. The inductance generates mainsborne and radio frequency interference which disrupts the triggering of the thyristor. The circuit shown in Figure 7.19 is essentially the same as Figure 7.18 except for the addition of a 1 kΩ resistor in the gate circuit. This is known as a 'bleed' resistor which makes the gate less sensitive to the interference generated by the inductive load and prevents unwanted triggering which allows more accurate speed control of the motor.

Varying the 5 kΩ resistor will vary the trigger angle of the thyristor. The 10 kΩ and 5 kΩ resistors act as a voltage divider which, in this case, causes approximately one third of the voltage available to

be dropped across the variable resistor. The diode D_2 will ensure that only a positive voltage is applied to the gate of the thyristor. Diode D_1 prevents the negative half cycle of the mains waveform being presented to the thyristor gate. The speed of the electric motor is controlled by the trigger action of the thyristor gate. However, it cannot run at full speed because the thyristor is a half-wave device, which cuts out the negative half cycle of the mains supply as shown in Figure 7.17. Closing the switch will cut out the thyristor connecting the mains supply to the motor, which will then run at its maximum speed. Thryistor speed control is only possible in this circuit with the switch open.

Flywheel diode

When highly inductive loads such as motor or discharge lighting circuits are to be switched in electrical installation work the IEE Regulations recommend that the functional switches be rated at *twice* the total steady current of the circuit. This is to avoid excessive wear of the switch contacts resulting from the back emf which is generated when inductive circuits are switched. (See Chapter 9 of Advanced Electrical Installation Work.)

A thyristor switches the load on and off during every half cycle and, when used to switch inductive loads, it has been found that the back emf causes the trigger circuit to lose control and, therefore, the current continues to flow in the load when the thyristor has attempted to switch off. This problem can be eliminated by fitting a *bypass* or *flywheel* diode which is connected in parallel with the load. This allows the reverse current generated by the back emf to circulate harmlessly around the loop formed by the load and the diode, leaving the thyristor to be triggered normally by the gate signal.

A flywheel diode is fitted to the motor circuit, shown in Figure 7.20. The variable resistor *R* varies the trigger angle and, therefore, the point at which the thyristor switches on during each positive half cycle. This varies the power available to the motor which varies the speed. The thyristor switches off at the end of each half cycle and the motor generates a back emf, which circulates a clockwise current around the parallel circuit made up of the motor and the flywheel diode 'D'. The

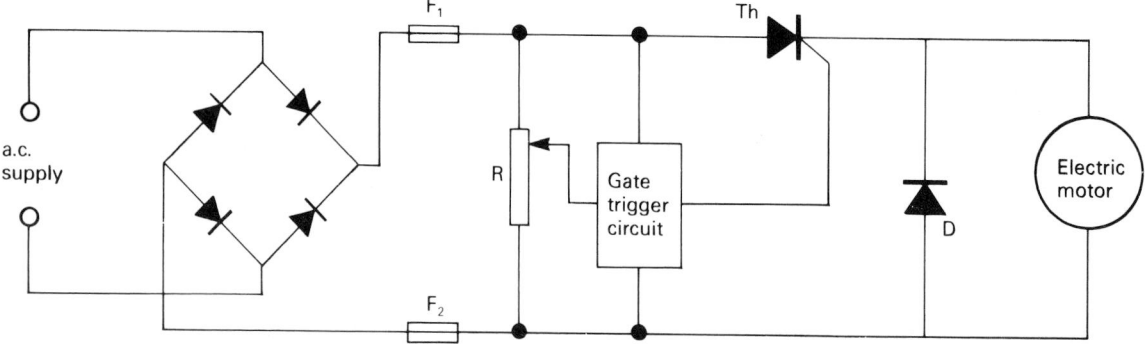

Figure 7.20 Thyristor speed control of an electric motor fitted with a flywheel diode

energy contained in the back emf is then dissipated harmlessly in the resistance of this parallel circuit.

If the motor fails to operate, the following simple checks can be made to the circuit.

1. Test for a satisfactory output voltage from the rectifier at fuses F1 and F2.
2. Connect a short piece of cable between the anode and cathode of the thyristor. This link will short out the thyristor and test the motor. If the motor runs, the fault is in *either* the thyristor or the gate trigger circuit. Remove the link.
3. Disconnect the gate trigger circuit connections to the variable resistor and the thyristor gate. Turn the variable resistor *R* to the middle position and connect a short piece of cable from the wiper connection of the variable resistor to the thyristor gate connection. If the motor operates, the fault is in the trigger unit. If the motor fails to operate, the thyristor is probably faulty and should be removed from the circuit and tested as described in Figure 6.23.

Three-phase power control using thyristors

A three-phase supply can be used to drive a d.c. motor if the circuit shown in Figure 7.21 is assembled. The thyristor trigger pulses are electronically separated by 120° so that only one thyristor conducts at any one time.

When TH1 is triggered, current passes to the load and returns via either D2 or D3, whichever diode is the most negative at each instant of time.

In practice current will return by D2 for a short period of time and then by D3. When TH2 is triggered 120° later, current passes to the load and returns via either D1 or D3. TH3 is then triggered and the current returns by D1 or D2. In this way, the load is presented with a unidirectional supply which contains only a small a.c. ripple.

Power control using triac

The thyristor is a half-wave device, allowing control of only half the available power at any one time, which is uneconomical. The triac is a single three-terminal device which is equivalent to two back-to-back thyristors. This allows it to conduct on both halves of the supply waveform and to be triggered on with either polarity of gate signal. Figure 7.22 shows a simple triac circuit and the resulting waveforms.

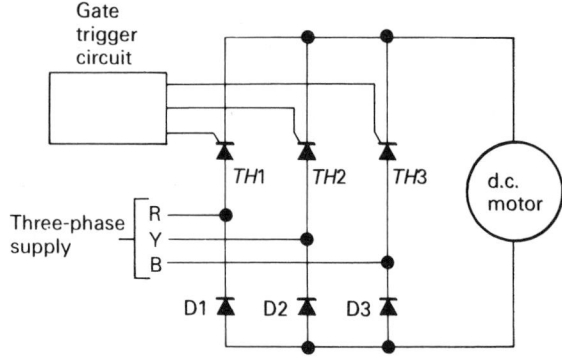

Figure 7.21 Three-phase power control using thyristors

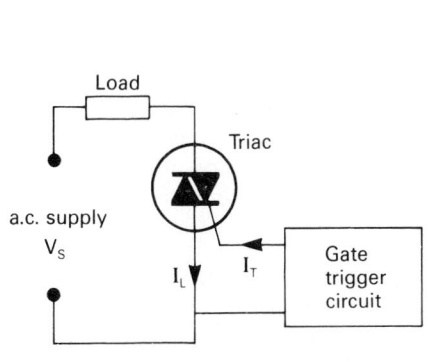

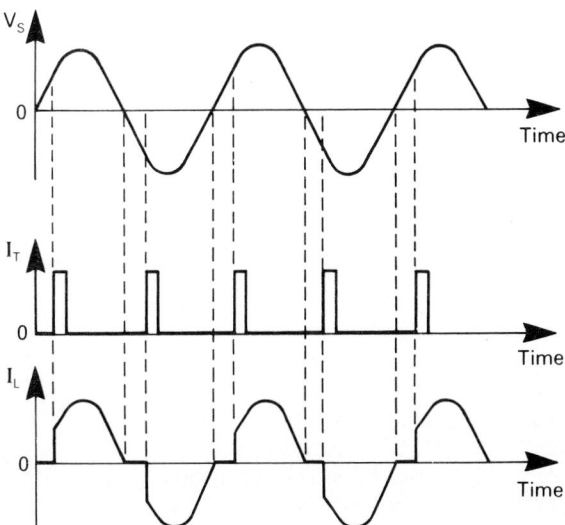

Figure 7.22 Simple triac circuit and waveforms

Diac as a trigger device

The diac is used mainly as a trigger device for thyristors and triacs. It is a two-terminal device which is equivalent to two back-to-back zener diodes. When a pre-determined voltage is reached, typically 30 V, the diac conducts providing the trigger pulse to the gate of the triac. The trigger pulse occurs on both positive and negative half cycles and so the triac can take advantage of all the power available.

Figure 7.23 shows a diac-triggered triac controlling a lamp dimmer circuit. When the capacitor C has charged up to the diac switching voltage, say 30 V, the diac will conduct, the capacitor will discharge through the diac, providing a trigger pulse to the triac, which will switch on and control the current flowing through the load and, therefore, the power to the lamp. Increasing the value of resistance R will increase the time taken for the capacitor to reach the diac switching voltage and, therefore, the point of switch on will be delayed. (See charging capacitors Chapter 3.) Reducing the value of resistance will reduce the time constant of the C-R circuit and the triac will switch on earlier in each half cycle, delivering more power to the lamp which will illuminate brightly. The waveforms will be similar to those shown in Figure 7.22 with the variable resistor controlling the trigger angle of the triac and, therefore, the lamp bright-

ness. The R-C circuit on the right of the circuit diagram and connected across the triac is known as a *snubber circuit*.

Snubber network

Triacs are also susceptible to false triggering caused by mainsborne interference spikes. The *snubber circuit* takes out these instantaneous changes and prevents the triac triggering inadvertently. When the triac is used with inductive loads, it is essential to connect a *snubber network* across the triac. Typical values are 100 Ω and 0.1 µF as shown in Figure 7.23.

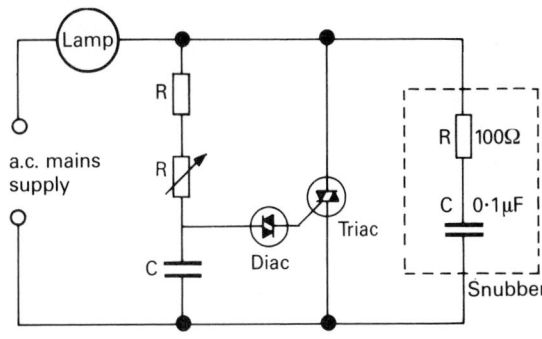

Figure 7.23 Lamp dimmer circuit using diac triggering

Manufacturers' data sheet information and practical information on thyristors and triacs is included in the Appendix.

Amplifiers

Amplification is one of the most important functions in electronics. Amplifiers are usually concerned with converting small changes of voltage at the input of the amplifier into larger changes of voltage at the output. They are used in hi-fi equipment to increase the signal strength from a tape, record or compact disc before it can be usefully played through the speaker system. The very small signal from a transducer such as a strain gauge can only be used for display purposes after being amplified as described in Chapter 9.

The amount by which an amplifier amplifies is known as the *gain* of the amplifier and has the general symbol A

$$\therefore \text{Gain} = A = \frac{\text{output signal}}{\text{input signal}}$$

In a voltage amplifier the amplified quantity is voltage and, therefore, the gain of the amplifier can be determined by

$$\text{voltage gain} = A_V = \frac{\text{output voltage}}{\text{input voltage}}$$

In a similar way

$$\text{current gain} = A_I = \frac{\text{output current}}{\text{input current}}$$

The feature of an amplifier which makes it important is that it can provide power gain and not just voltage gain. If we multiply the voltage gain by the current gain we obtain the power gain.

$$\therefore \text{ power gain} = A_P = \frac{\text{output power}}{\text{input power}}$$

$$= A_V \times A_I$$

In electronics, as in life, we never get something for nothing and this power gain must come from another energy source. In practice, every electronic amplifier must be supplied with a constant voltage from either a battery or a d.c. stabilised power supply circuit such as those described earlier in this chapter.

In many practical circuits the required voltage gain cannot be provided by any one amplifier stage. In this case the separate amplifier stages are connected together in series or cascade, output to input, and the total voltage gain can then be found by multiplying together the individual voltage gains. The total voltage gain of a number of stages is given by

$$\text{Total gain } A_V = A_{V_1} \times A_{V_2} \times A_{V_3} \text{ etc.}$$

Operational amplifier

The actual amplifier circuit can be made up from individual discrete transistors, or the most complex arrangement of transistors, resistors and capacitors can be integrated on to a tiny silicon chip which is then known as an integrated circuit or IC. When an amplifier is made in this way it is called an operational amplifier or op amp. The circuit symbol and appearance are shown in Figure 7.24. This is the easy way to amplify weak signals and although many different types of op amp are available, the 741 is the industry standard. The gain of a 741 op amp is about 100,000 compared with approximately 100 for a transistor amplifier. For most practical purposes a gain of this magnitude is much too large and unstable but the actual performance can be modified with external components. This is found to reduce the distortion of the output signal, increase the range of frequencies which can be amplified and reduce but stabilise the gain of the amplifier. These advantages outweigh the loss of gain which can easily be increased by using two or more op amp stages.

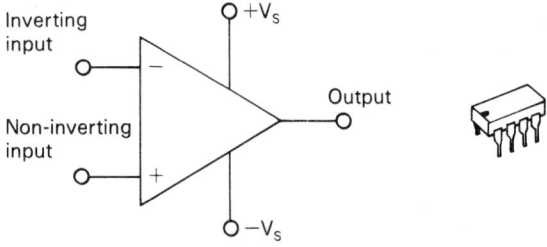

Figure 7.24 Circuit symbol and appearance of a 741 OpAmp

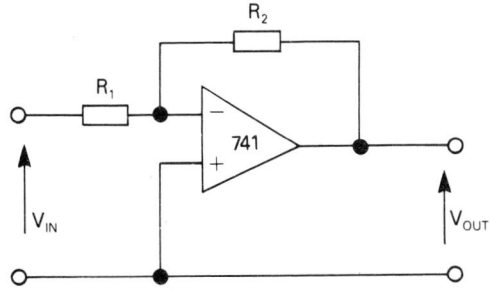

Figure 7.25 The basic OpAmp circuit

The basic circuit diagram for an op amp voltage amplifier is shown in Figure 7.25. The output voltage of this op amp will be greater than the input voltage by a factor determined by gain A_v and in antiphase to it, because the input is connected to the inverting (-) terminal and negative feedback is provided by R_2. The gain is controlled by the ratio of R_2/R_1 and in the case of negative feedback, we add a negative sign to the formulae to obtain the equation

$$A_V = \frac{-R_2}{R_1}$$

Therefore, the output V_o of Figure 7.25 will be

$$V_O = -V_{in}\frac{R_2}{R_1} \text{ volts}$$

If the value of R_2 and R_1 in Figure 7.25 were 200 kΩ and 1 kΩ, the gain of the op amp would be 200. With an input signal Vin of 1 µV, the output would be

$$V_O = -1 \times 10^{-6} \times \frac{200 \text{ k}\Omega}{1 \text{ k}\Omega} = -0.2 \text{ mV}$$

That is, the output is 200 times greater than the input and inverted. To an electrician 0.2 mV still seems an insignificant voltage but a voltage gain of 200 is significant and these values are what we could expect from a strain gauge circuit or the pick-up of a compact disc player.

When an amplifier is used to amplify the input voltage or current in such a way that the output is an enlarged copy of the input and is not distorted, it is said to be a small signal amplifier. When an amplifier is used to amplify the power of an input

signal it is said to be a power amplifier. Figure 7.26 shows the circuit diagram of an audio frequency amplifier. The left-hand side of the circuit, the op amp, is a small signal voltage amplifier which is used to amplify a small signal from, for example, the ear piece jack plug of a tape recorder. The right-hand side of the circuit is the power amplifier which is required to drive the speaker. This is made from a pair of complementary power transistors, one is an npn and the other a pnp transistor which have been *matched* so that they have the same gain and other properties. When the voltage on the top transistor is positive the voltage on the bottom transistor is negative and vice versa. The amplification of each half of the voltage waveform is, therefore, shared between the two transistors. A circuit which is constructed in this way is known as a push-pull amplifier. The additional power required to drive the speaker in this circuit comes from the 9 V batteries.

Bandwidth of an amplifier

The bandwidth of an amplifier is the range of frequencies within which the gain does not fall below about 0.7 of the maximum gain. The points at which this happens, f_1 and f_2 in Figure 7.27, are called the 3 dB points. The decibel (dB) scale is used in electronics to compare signal power levels or the loudness of a sound. The unit is named in honour of Alexander Graham Bell (1847–1922), the pioneer of the telephone.

Most audio amplifiers, such as the one shown in Figure 7.26, are designed to amplify the range of frequencies to which the human ear will respond. This is approximately 15 Hz to 15 kHz although some manufacturers of good quality audio amplifiers design their circuits with an upper frequency limit of about 40 kHz. Amplifying the frequencies well above the range of human hearing can improve the sound quality within the audio frequency range.

The relationship between gain and the frequency range of a 741 op amp is shown in Figure 7.28. Reducing the gain increases the frequency range. If the circuit was designed to have a gain of say 1000, the amplifier will respond to frequencies in the range 1 Hz to 1 kHz. However, this upper frequency limit is too low if the circuit is to be used as an audio amplifier. The gain will, therefore, need to be reduced to between 10 and 15 to obtain

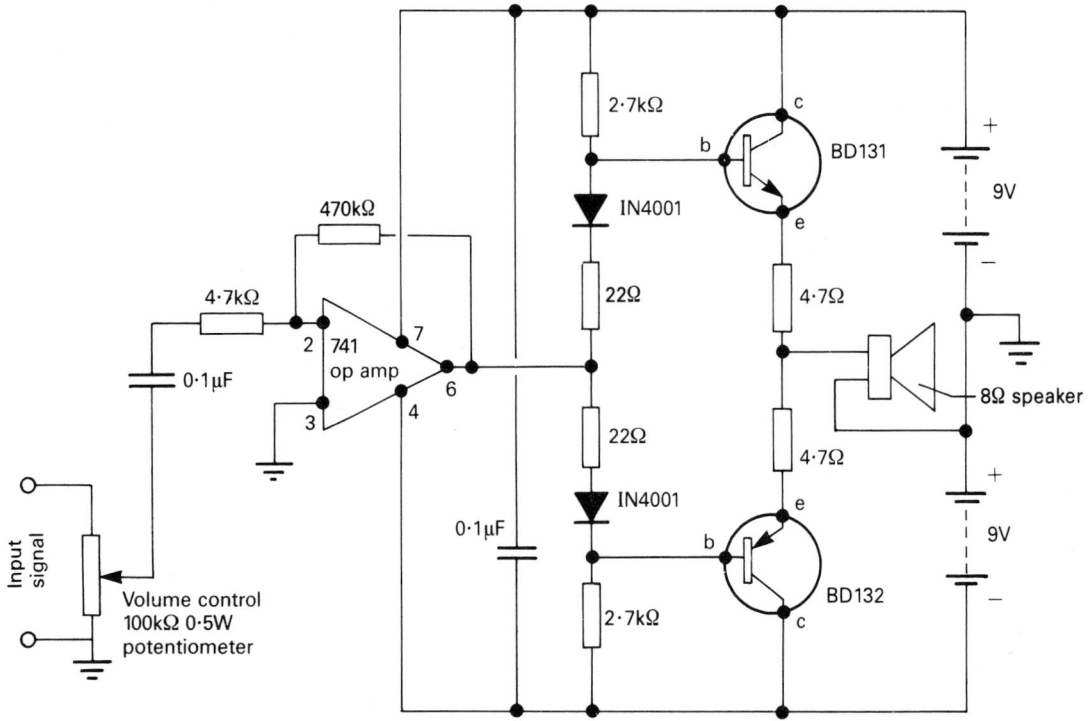

Figure 7.26 An audio frequency amplifier

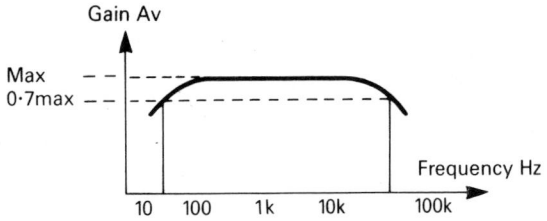

Figure 7.27 Amplifier gain/frequency response curve

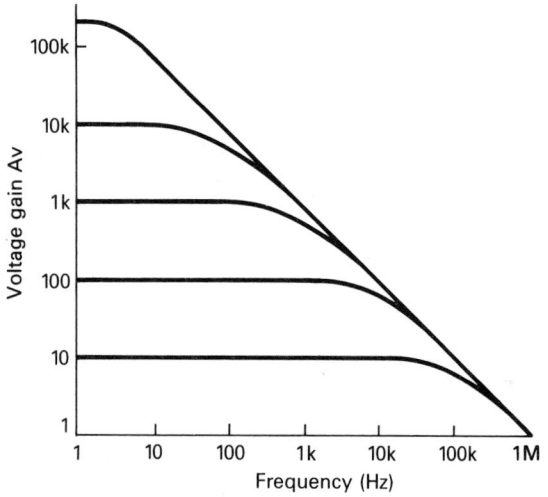

Figure 7.28 Frequency response of a 741 OpAmp for various values of gain

an upper frequency limit of about 20 kHz if the circuit is to be used as an audio amplifier.

Filters

Filters are circuits which allow some a.c. frequencies to pass through them more easily than others. There are many instances in electronics where the frequency bandwidth of a circuit must be restricted, for example, to reduce the signal strength-

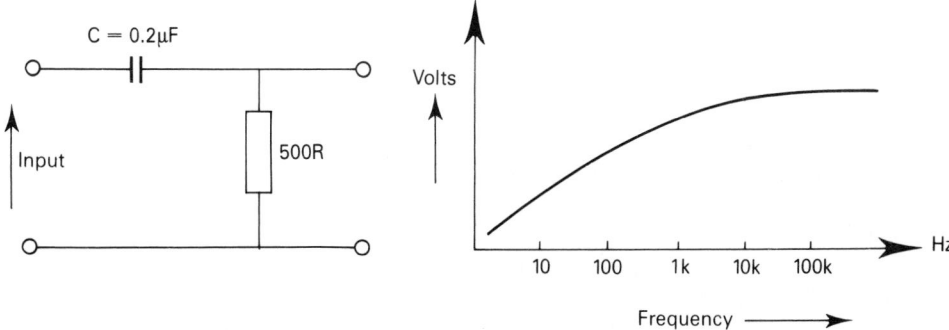

Figure 7.29 High pass filter

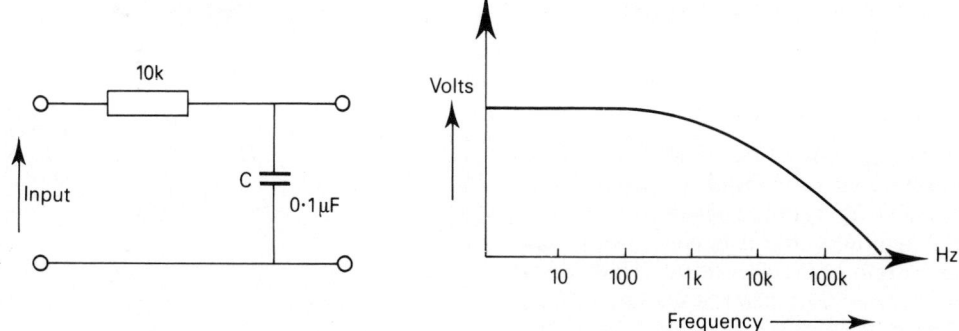

Figure 7.30 Low pass filter

to-noise ratio which is directly proportional to bandwidth.

If an old 78 rpm gramophone record is played on a modern wide frequency range hi-fi system, it may be difficult to hear the music above the surface noise. However, the same record played on a narrow frequency range 1950 vintage gramophone may sound quite acceptable. This is because the bandwidth is wide enough to pass the music but narrow enough to limit the noise frequencies. Two of the most simple filters are called *low pass* and *high pass* filters.

Figure 7.29 shows a simple first-order high pass filter which is made up of a capacitor and resistor. The reactance (resistance to a.c.) of a capacitor reduces as the frequency increases and, therefore, at low frequencies the reactance will be high and at high frequencies the reactance will be low. This circuit will, therefore, allow high frequencies to pass more easily than low frequencies and hence, is

called a high pass filter. High pass filters can be used to block d.c. when a number of a.c. amplifier stages are connected together.

Figure 7.30 shows a simple first-order low pass filter. The resistor and capacitor are connected the opposite way round and the circuit will, therefore, pass low frequencies better than high frequencies, hence it is called a low pass filter. Low pass filters can be used to block a.c. ripple on d.c. power supplies as shown in Figure 7.9, which shows a slightly improved low pass filter arranged in a π configuration.

Testing audio amplifiers

If the audio amplifier circuit shown in Figure 7.26 is assembled on a prototype board as shown in Figure 2.21, a number of tests can be carried out.

The circuit must be built *exactly* as shown in the

circuit diagram but a 10 Ω 1 watt resistor is used as the load in place of the 8 Ω speaker. The component pin connection information given in Figure 7.31 will help with the actual construction.

Adjust the settings on a double output power supply unit (PSU) to 9 V and connect with flying leads to the circuit assembled on the prototype board, making sure that the polarities are as shown in the circuit diagram.

Connect a signal generator to the amplifier input terminals and adjust it to give a sine wave at 1 kHz which is the industrial standard test frequency for audio amplifiers. Adjust the volume control to the maximum position. Set all the CRO controls to the calibrate position and connect the leads across the amplifier output terminals, that is across the 10 Ω load resistor. You are now ready to carry out the tests.

Testing for signal distortion

Increase the voltage output from the signal generator while observing the amplifier output waveform on the CRO. You should observe a sine wave increase in magnitude until it begins to distort at the top and bottom of the waveform. That is, the sine wave will *clip* or flatten at the top and bottom of the waveform. Reduce the input until the output waveform is once more sinusoidal and measure the input voltage being delivered by the signal generator. This is the maximum value of input voltage which can be applied to the amplifier before distortion occurs.

Testing for gain

Adjust the voltage output from the signal generator to a value of about 20 mV peak to peak at a frequency of 1 kHz. Measure the output and input voltage with the CRO and calculate the amplifier gain by dividing the output voltage value by the input voltage value.

Measuring the bandwidth

To be suitable as an audio amplifier the circuit shown in Figure 7.26 must amplify most of the frequencies to which the human ear will respond, that is between about 15 Hz and 15 kHz. To measure the actual bandwidth of the amplifier we must, therefore, make a series of gain measurements over this range of frequencies in order to plot a

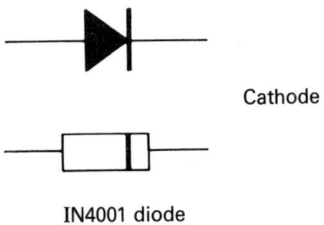

Cathode

IN4001 diode

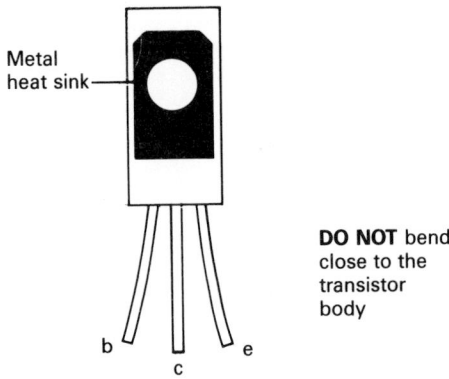

Metal heat sink

DO NOT bend close to the transistor body

b   e
c

Pin identification of both transistors

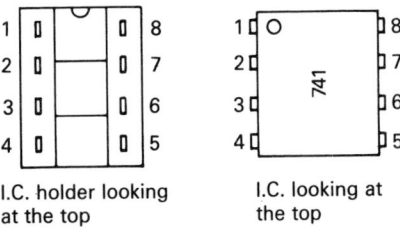

I.C. holder looking at the top   I.C. looking at the top

Pin identification of 741 I.C.

Figure 7.31 Component pin connection information to be used when building the audio amplifier circuit shown in Fig.7.26

gain/frequency response curve such as that shown in Figure 7.27. To do this we must vary the output frequency while maintaining a constant output voltage from a test signal source such as a signal generator as follows.

Set the output frequency of the signal generator to 10 Hz and adjust the voltage level to 20 mV peak to peak. Use the CRO to measure the voltage

Table 7.1 A table of results which is suitable for measuring the bandwidth of an audio amplifier

Frequency Hz	V_{in}	V_{out}	Gain $= Av = V_{out}/V_{in}$
10	20 mV		
30	20 mV		
60	20 mV		
100	20 mV		
500	20 mV		
1,000	20 mV		
4,000	20 mV		
6,000	20 mV		
8,000	20 mV		
10,000	20 mV		
20,000	20 mV		

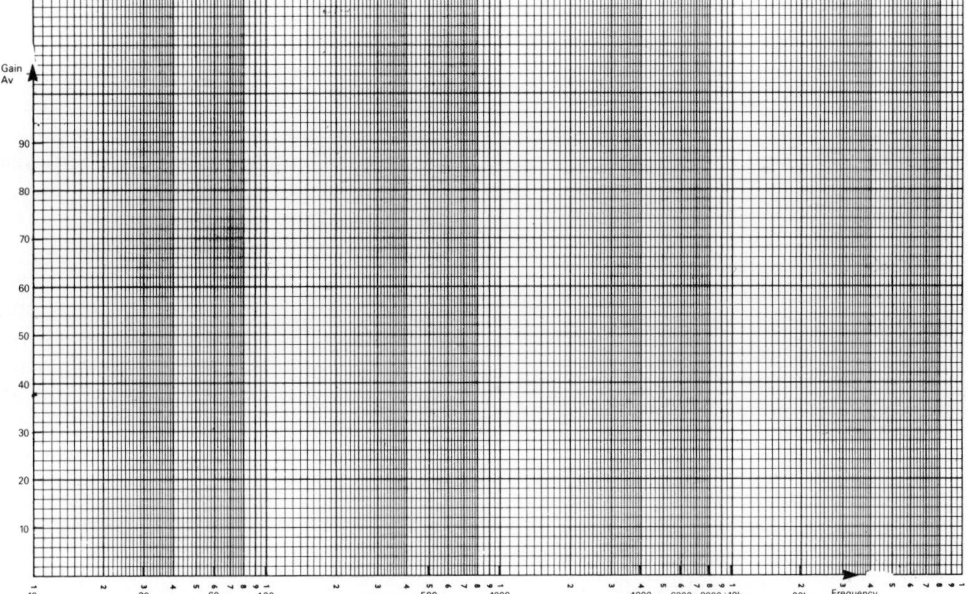

Figure 7.32 Graph paper for plotting the frequency response curve of an audio amplifier

at the input and output terminals of the amplifier. The gain at this frequency can then be calculated as before and the results entered in Table 7.1. Next, set the output frequency of the signal generator to 30 Hz and adjust the voltage to 20 mV peak to peak. Use the CRO to measure the voltage at the input and output terminals of the amplifier and once more calculate the gain and enter the results

in Table 7.1. Continue to take and enter readings in this way until Table 7.1 is completed.

The gain/frequency response curve can then be plotted on the Log/Lin graph paper as shown in Figure 7.32 using the gain and frequency values tabulated in Table 7.1. The logarithmic horizontal scale is necessary to accommodate the large frequency range of an audio amplifier. When all the points have been plotted they can be joined to form a bandwidth curve which should be similar to that shown in Figure 7.27. The bandwidth is the range of frequencies enclosed by the upper and lower frequency points when the gain is equal to 0.7 of the maximum gain. What is the bandwidth of your amplifier? The bandwidth of an expensive music centre amplifier might be between 20 Hz and 40 kHz while the telephone system is found to give acceptable voice reproduction with a bandwidth from 300 Hz to 3.4 kHz. Do you think that your amplifier will give acceptable results? You can test it by playing a tape recorder through the amplifier. To do this, remove the 10 Ω load resistor used for testing and connect an 8 Ω speaker across the load. Connect the 9 V supplies from the double PSU as described earlier and connect to the input signal connections of the amplifier, a connection from the ear phones jack plug of a tape recorder. The ear phones signal can in this way be played through the amplifier which you have constructed.

Waveforms

The electricity generating stations generate a sinusoidal alternating voltage waveform from an alternator which is usually driven by a steam turbine. A sinusoidal waveform is a *repetitive*, *analogue* waveform. Repetitive because in the case of the a.c. mains it repeats itself fifty times each second, and analogue because it varies smoothly and continuously between two extremes. In everyday life, analogue systems are all around us. Television, radio and telephone signals are all analogue, the amplifier considered earlier is an analogue amplifier. A car speedometer and fuel gauge are analogue instruments. Analogue electronics is one of *two* main branches of electronics, the other is *digital electronics*. Digital signals do not change smoothly and continuously between voltage levels but have two quite definite levels, either on or off. Digital signals are in the form of electrical pulses whose outputs involve only two levels of voltage, called high or low, where high might be +5 V and low 0 V. The term *mark-to-space* ratio is used in connection with digital signals, particularly square, rectangular and pulse waveforms, and is given by

$$\text{mark-to-space ratio} = \frac{\text{mark time}}{\text{space time}}$$

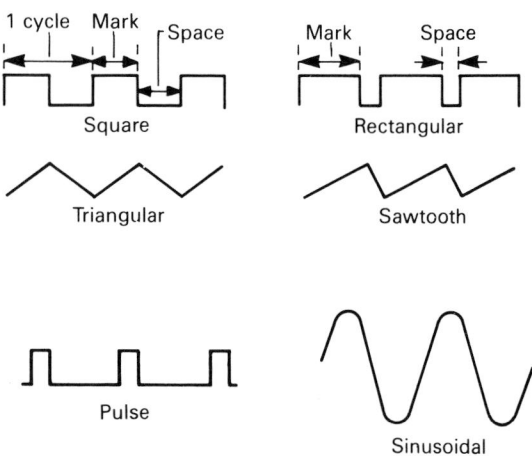

Figure 7.33 Some common waveforms

For a square wave the mark-to-space ratio will be 1 because the mark time and space times are equal. Digital circuits are used in pocket calculators, electronic watches, domestic appliances and motor car control systems and increasingly in communication systems.

In electrical installation work we are mostly concerned with sinusoidal voltage waveforms but in electronics other wave shapes are of great importance. For example, square waves are used for timing and oscillator circuits, sawtooth waveforms for the CRO time base and pulses to trigger a thyristor or triac. Figure 7.33 shows some common waveforms.

Sawtooth waveform generator

If a circuit is constructed as shown in Figure 7.34 the capacitor will charge up at a rate which is determined by the values of C and R when the supply is switched on. (See Chapter 3, *charging capacitors* for the relevant theory.) However, if a neon lamp or a thyristor is connected across the capacitor so that the capacitor is discharged quickly at some predetermined point during the lower linear part of the capacitor growth curve, then a sawtooth waveform is generated which has a relatively long rise time and a short discharge time. When the capacitor is discharged to zero volts the neon lamp or thyristor will switch off and the capacitor will once more begin to charge up to the predetermined discharge point and the sawtooth

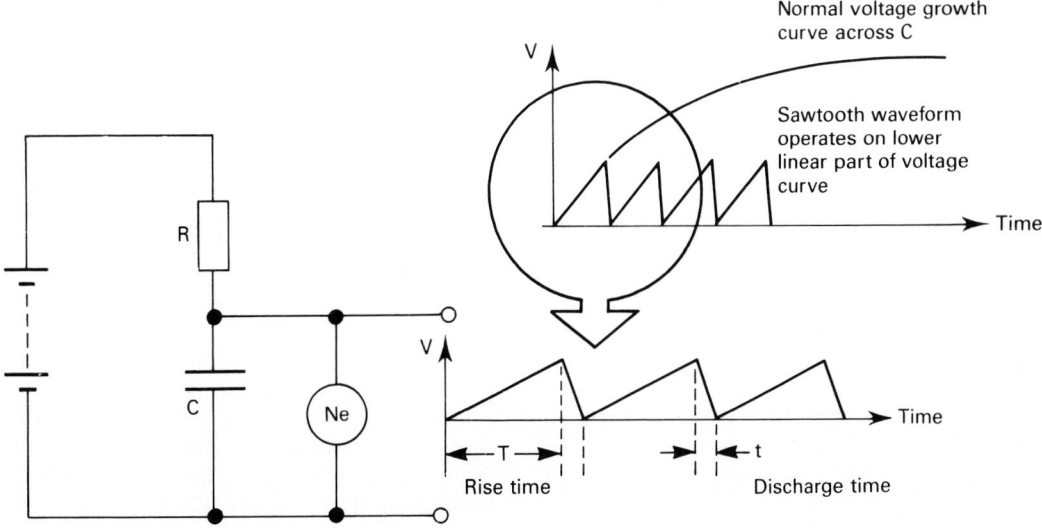

Figure 7.34 Sawtooth waveform generator circuit

waveform will, therefore, be generated continuously. The long rise time to short discharge time is characteristic of a sawtooth waveform. A sawtooth waveform generator is used to drive the electron beam across the X axis of the CRO screen. During the long rise time the beam sweeps from the left- to the right-hand side of the screen. Flyback occurs during the short discharge period placing the beam once more on the left-hand side ready for a further sweep across the screen.

Transistor switching using a capacitor

This is another application of the circuit theory associated with *charging capacitors* discussed in Chapter 3. If a circuit is constructed as shown in Figure 7.35, the lamp will illuminate when the supply is switched on because a signal will be applied to the base region of the transistor, but as the capacitor charges up, the voltage across the resistor will fall to zero, switching off the signal to the transistor base, which will switch off the transistor and the signal lamp. The capacitor charging current can be used to switch the transistor on or off at a rate which is dependent upon the value of the C-R time constant. This principle can be applied to the electronic circuits known as multivibrators.

Multivibrators

Multivibrators are a class of electronic switching circuits which are also known as *relaxation oscillators* because their operation is characterised by a period of vigorous activity followed by a period of

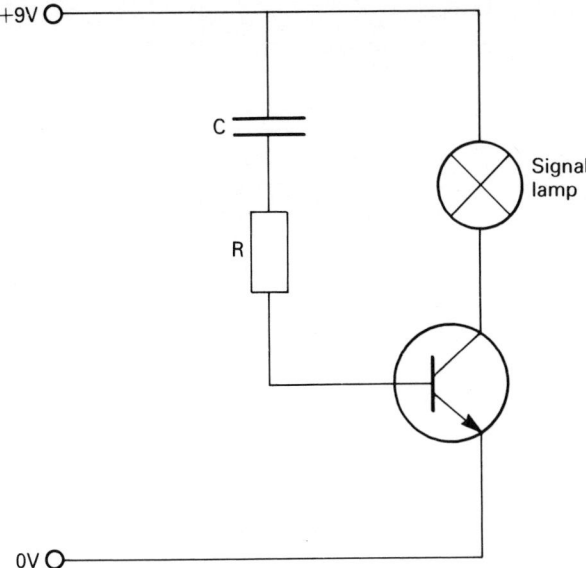

Figure 7.35 Transistor switching using a capacitor

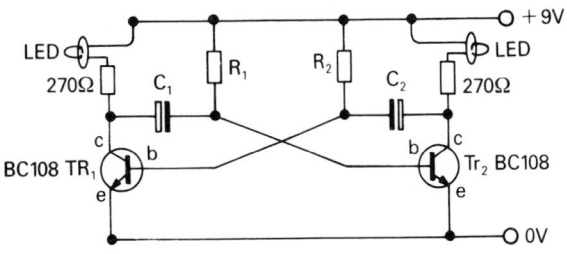

Figure 7.36 Astable multivibrator

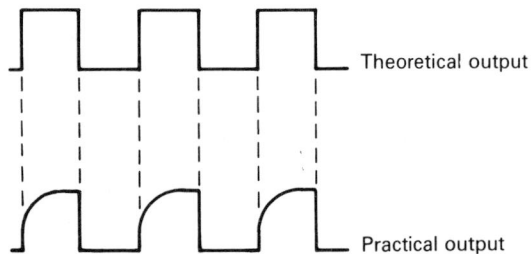

Figure 7.37 Astable multivibrator output

relaxation. They produce an output which is generally a series of square or rectangular pulses by switching between two transistors. Figure 7.36 shows the circuit diagram for an *astable* multivibrator. Astable means not stable or free running and, therefore, the circuit generates a continuing train

of pulses at the output. The two transistors are coupled together, so that while one transistor is switched on, the other is switched off. Each transistor then switches automatically to its other state and, as a result, the output voltage which can be taken from the collector of either transistor is alternatively *high* (+8 V) or *low* (almost 0 V) generating a series of theoretically square pulses. In practice, the switch over from one transistor to another is not instantaneous and, therefore, the output will not be perfectly square as shown by Figure 7.37. The length of time each transistor is switched on depends upon how long it takes the capacitors to charge up through the resistors. TR_1 is switched by R_2 C_2 and TR_2 by R_1 C_1. If $C_1 = C_2$ and $R_1 = R_2$ the 'off' and 'on' times for each transistor will be equal and a square wave will be generated. The *rate* at which the signal lamps will flash can be varied by changing the capacitor and resistor values as suggested by Table 7.2. Reducing the values of C and R will reduce the circuit time constant and increase the lamp flashing rate. Using different values of C_1 and C_2 or different values of R_1 and R_2 will vary the mark-to-space ratio and a rectangular waveform will be generated which will result in an uneven flashing rate of the two signal lamps.

It may help your understanding of the circuit if you were to assemble the multivibrator circuit on an S-DeC; vary the component values as suggested by Table 7.2 and observe the results.

Table 7.2   A results table for the multivibrator circuit investigation

C_1, C_2	220 μF	220 μF	220 μF	22 μF
R_1, R_2	4.7 kΩ	10 kΩ	22 kΩ	10 kΩ
Number of flashes per minute				
Time constant				

CHAPTER 8

Logic gates and digital electronics

Digital electronics embraces all of today's computer-based systems. These are decisionmaking circuits which use what is known as combinational logic in applications such as industrial robots, industrial hydraulic and pneumatic systems (PLCs), telephone exchanges, motor vehicle and domestic appliance control systems, children's toys and their parents' personal computers and audio equipment. Digital electronics is concerned with straightforward two-state switching circuits. The simplicity and reliability of this semiconductor transistor switching has encouraged designers to look for new digital markets. Traditional applications which have analogue inputs, such as audio recordings, are now using digital techniques, with the development of analogue-to-digital converters. These convert the analogue voltage signals into digital numbers. A digital and analogue waveform are shown in Figure 8.1.

The digital waveform has two quite definite states, either on or off, and changes between these two states very rapidly. An analogue waveform changes its value smoothly and progressively between two extremes.

In an analogue system, changes in component values due to ageing and temperature can affect the circuit's performance. Digital systems are very much less susceptible to individual component changes. Another significant advantage of digital circuits is their immunity to noise and interference signals. With analogue circuitry this is a nuisance, particularly when signal levels are very small and, therefore, easily contaminated by noise. Digital signals, however, have a very large amplitude and can, therefore, be made relatively free of noise which helps manufacturers to achieve a very high quality of sound reproduction, as anyone who listens to a compact disc recording can testify. Logic circuits have been developed to deal with these digital, two-state switching circuits. Information is expressed as *binary numbers*, that is numbers which consist of ones and zeros. These two binary states are represented by low and high voltages, where low voltage is 0 V and high voltage is say +5 V. The low level is called Logic 0 and the high level Logic 1. When the voltage level of a digital signal is not rapidly changing it remains steady at one of these two levels. Information is processed according to rules built into circuits made up of single units called *Logic gates*. Each unit is called a *gate*, because like a gate it can allow information to pass through or stop it, and *Logic*

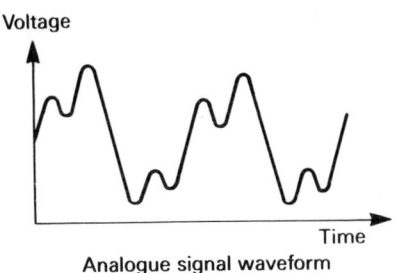

Analogue signal waveform

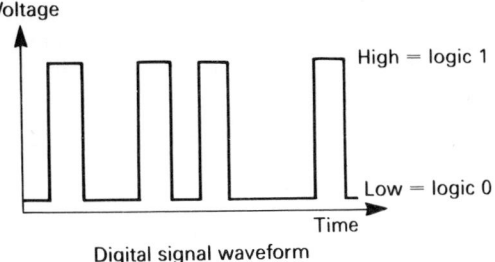

Digital signal waveform

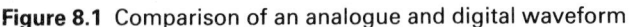

Figure 8.1 Comparison of an analogue and digital waveform

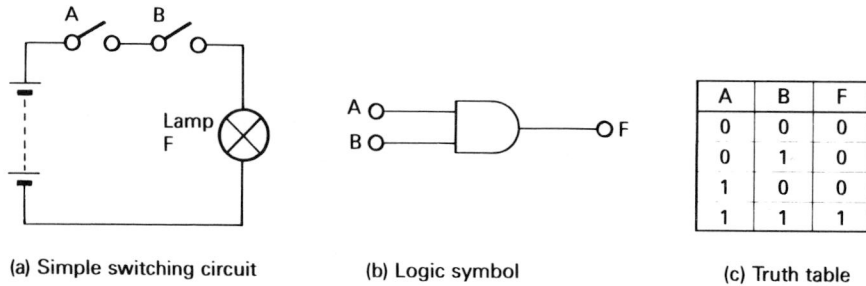

(a) Simple switching circuit   (b) Logic symbol   (c) Truth table

A	B	F
0	0	0
0	1	0
1	0	0
1	1	1

Figure 8.2 The AND gate

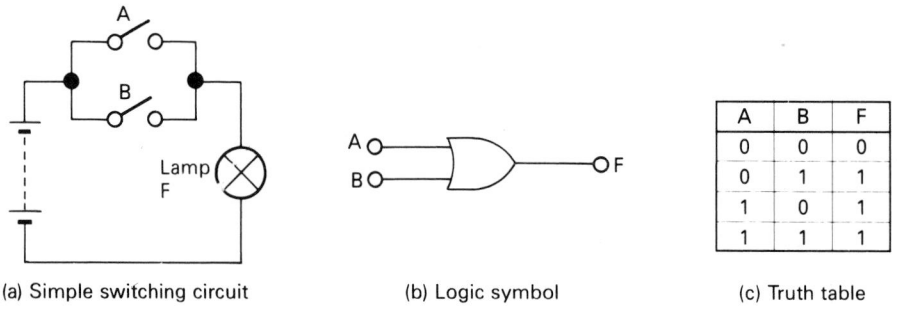

(a) Simple switching circuit   (b) Logic symbol   (c) Truth table

A	B	F
0	0	0
0	1	1
1	0	1
1	1	1

Figure 8.3 The OR gate

gate because it behaves according to rules which can be described by logical or predictable statements. A logic gate may have a number of inputs but has only one output which can only be either logic 1 or logic 0, no other value exists. The basic range of logic gates is known by the names AND, OR, NOT, NOR and NAND.

The AND logic gate

The operation of this gate can probably best be understood by drawing a simple switch equivalent circuit, as shown in Figure 8.2. The logic symbol is also shown. The signal lamp will only illuminate if switch A *and* switch B are closed, or we could say the output F of the gate will only be at logic 1 if input A *and* input B are both at logic 1.

If the AND gate was operating a car handbrake warning lamp, it would only illuminate when the handbrake *and* the ignition was on. The *truth table* shows the output state for all possible combinations of inputs.

The OR gate

The OR gate can be represented by parallel connected switches, as shown in Figure 8.3 which also shows the logic symbol. In this case the signal lamp will only illuminate if switch A *or* switch B *or* both switches are closed. Alternatively, we could say that the output F will only be at logic 1 if input A *or* input B *or* both inputs are at logic 1.

If the OR gate was operating an interior light in a motor car, it would illuminate when the nearside door was opened *or* the offside door was opened *or* when both doors were opened. The truth table shows the output state for all possible combinations of inputs.

The Exclusive -OR gate

The Exclusive -OR gate is an OR gate with only two inputs which will give a logic 1 output only if input A *or* input B is at logic 1, but *not* when both

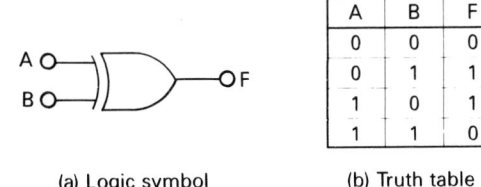

A	B	F
0	0	0
0	1	1
1	0	1
1	1	0

(a) Logic symbol   (b) Truth table

Figure 8.4 The Exclusive -OR gate

A and B are at logic 1. The symbol and truth table are given in Figure 8.4.

The NOT gate

The NOT gate is a single input gate which gives an output that is the opposite of the input. For this reason it is sometimes called an *inverter* or a *negator* or simply a *sign changer*. If the input is A, the output is *not* A which is written as Ā (bar A). The small circle on the output of the gate always indicates a change of sign.

If the NOT gate was operating a spin dryer motor it would only allow the motor to run when the lid was *not* open. The truth table shows the output state for all possible inputs in Figure 8.5.

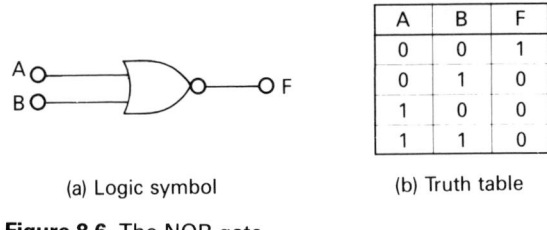

INPUT	OUTPUT
0	1
1	0

(a) Logic symbol   (b) Truth table

Figure 8.5 The NOT gate

The NOR gate

The NOR gate is a NOT gate and an OR gate combined to form a NOT-OR gate. The output of the NOR gate is the opposite of the OR gate as can be seen by comparing the truth table for the NOR gate in Figure 8.6 with that of the OR gate.

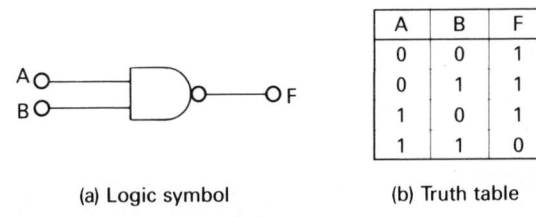

A	B	F
0	0	1
0	1	0
1	0	0
1	1	0

(a) Logic symbol   (b) Truth table

Figure 8.6 The NOR gate

The NAND gate

The NAND gate is a NOT gate and an AND gate combined to form a NOT-AND gate. The output of the NAND gate is the opposite of the AND gate as can be seen by comparing the truth table for the NAND gate in Figure 8.7 with that of the AND gate.

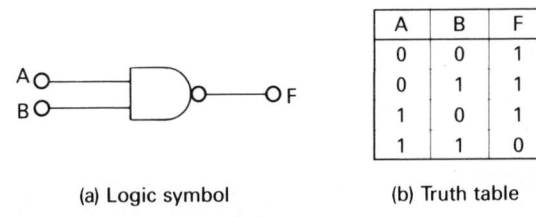

A	B	F
0	0	1
0	1	1
1	0	1
1	1	0

(a) Logic symbol   (b) Truth table

Figure 8.7 The NAND gate

Buffers

The simplest of all logic devices is the buffer. This device has only one input and one output, and its logical output is exactly the same as its logical input. Given that this device has no effect upon the logic levels within a circuit, you may be wondering what the purpose of such an apparently redundant device might be! Well, although the input and output voltage levels of the buffer are identical,

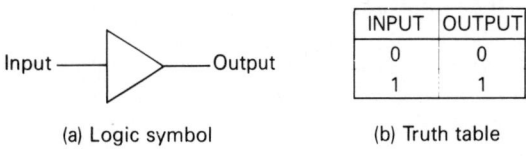

INPUT	OUTPUT
0	0
1	1

(a) Logic symbol   (b) Truth table

Figure 8.8 The Buffer

the *currents* present at the input and output can be *very* different. The output current can be much greater than the input current and, therefore, buffers can be said to exhibit *current gain*. In this way, buffers can be used to interface logic circuits to other circuits which demand more current than could be supplied by an unbuffered logic circuit. The symbol used to represent a buffer is shown in Figure 8.8.

Logic networks

Individual logic gates may be interconnected to provide any desired output. The results of any combination can be found by working through each individual gate in the combination or logic system in turn, and producing the truth table for the particular network. It can also be very instructive to build up logic gate combinations on a logic simulator and to confirm the theoretical results. This facility will undoubtedly be available if the course of study is being undertaken at a Technical college, Training Centre or Evening Institute.

Example 1

Two logic gates are connected together as shown in Figure 8.9. Complete the truth table for this particular logic network. In considering Figure 8.9 and working as always from left to right, we can see that an AND gate feeds a NOT gate. The whole network has two inputs, A and B, and one output F. The first step in constructing the truth table for the combined logic gates is to label the outputs of *all* the gates and prepare a blank truth table as shown in Figure 8.9. Let us call the output of the AND gate C (it could be any letter except A, B or F) and work our way progressively through the individual gates from left to right. For any two input logic gate, there are four possible combinations, 0 0, 0 1, 1 0 and 1 1. When these are included on the truth table it will appear as shown in Figure 8.10. The next step is to complete column C. Now, C is the output of an AND gate and can, therefore, only be at logic 1 when both A *and* B are logic 1. The truth table can, therefore, be completed as shown in Figure 8.11. The final step is to complete column F, the output of a NOT gate whose input combinations are given by column C. A NOT gate is a single input gate whose output is the opposite

A	B	C	F
0	0		
0	1		
1	0		
1	1		

Figure 8.10 Truth table for Example 1

A	B	C	F
0	0	0	
0	1	0	
1	0	0	
1	1	1	

Figure 8.11 Truth table for Example 1

A	B	C	D
0	0	0	1
0	1	0	1
1	0	0	1
1	1	1	0

Figure 8.12 The completed truth table for Example 1

of the input and, therefore, the output column F must be the opposite of column C, as shown by Figure 8.12. The truth table tells us that this particular combination of gates will give a logic 1 output with any input combination *except* when A and B are both at logic 1. This combination, therefore, behaves like a NAND gate as can be confirmed by referring to Figure 8.7.

Example 2

A NAND and NOT gate are connected together as shown in Figure 8.13. Complete a truth table for this particular network. The truth table for this particular combination can be constructed in exactly the same way as for Example 1. The NAND gate has two inputs P and Q and an output R. The NOT gate has an input R and output S.

All possible combinations of inputs are shown in columns P and Q of the truth table shown in Figure 8.14. A NAND gate will give a logic 1 output for *all*

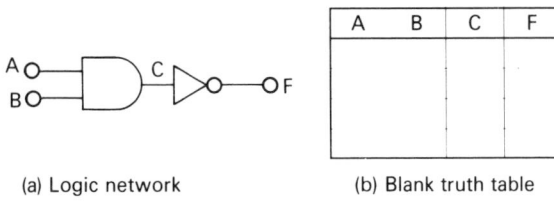

(a) Logic network   (b) Blank truth table

A	B	C	F

Figure 8.9 Logic network and blank truth table for Example 1

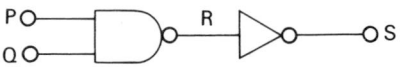

Figure 8.13 Logic network for Example 2

P	Q	R	S
0	0	1	0
0	1	1	0
1	0	1	0
1	1	0	1

Figure 8.14 Truth table for Example 2

combinations of inputs *except* when input A *and* B are at logic 1 as shown by column R of Figure 8.14. The second and final gate in this network is a NOT gate which provides an output which is the reverse of the input. The output, given by column S of the truth table, will therefore be the reverse of column R as shown in Figure 8.14.

This particular combination will, therefore, give a logic 1 output only when input P *and* input Q are at logic 1. Therefore, it can be seen that the combination of a NAND and a NOT gate produces the equivalent of an AND gate. This can be checked by referring back to Figure 8.2.

Example 3

A NAND gate has a NOT gate on each of its inputs as shown in Figure 8.15. Construct a truth table for this particular network. The NOT gates will invert or reverse the input. We can, therefore, call the output of these NOT gates, not A and not B, which is written as Ā and B̄. This then provides the input

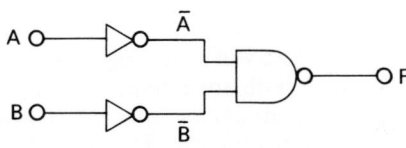

Figure 8.15 Logic network for Example 3

A	B	Ā	B̄	F
0	0	1	1	0
0	1	1	0	1
1	0	0	1	1
1	1	0	0	1

Figure 8.16 Truth table for Example 3

to the NAND gate. A NAND gate will provide a logic 1 output for any input combination *except* when both inputs are at logic 1. The truth table can, therefore, be developed as shown in Figure 8.16. It can be seen by referring back to Figure 8.3, and comparing the inputs A and B and output F, that this combination gives the network equivalent of an OR gate. That is, the output is at logic 1 if the input A *or* input B *or* both are at logic 1.

Example 4

A logic network is assembled as shown in Figure 8.17. Develop a truth table and describe in a sentence the relationship between the input and output. The truth table for this particular combination can be drawn up as shown in Figure 8.18. There are only two inputs A and B. The output C of the AND gate and the output D of the OR gate provide the input to a NOR gate, which provides the output F.

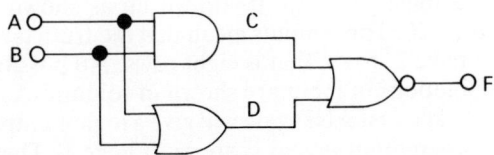

Figure 8.17 Logic network for Example 4

A	B	C	D	F
0	0	0	0	1
0	1	0	1	0
1	0	0	1	0
1	1	1	1	0

Figure 8.18 Truth table for Example 4

The output of an AND gate is high, that is at logic 1, only when input A *and* input B are at logic 1. Column C of the truth table shows the output of the AND gate for all combinations of input. The output of an OR gate is high, when input A *or* input B *or* both are high. This is shown by column D of the truth table. The input to the final NOR gate is provided by the logic levels indicated in columns C and D and the output F is, therefore, as shown in column F. The output of this combination of logic gates is high, that is at logic 1, only when input A *and* input B are low. This is equivalent to a single NOR gate.

In the examples considered until now, the inputs have been restricted to only two variables. In practice, logic gates may be constructed with many inputs and the truth tables developed as shown above. However, when there are more than three inputs the truth table becomes very cumbersome because the number of lines required for the truth table follows the law of 2^n where n is equal to the number of inputs. Therefore, a two-input gate requires 2^2 (4) lines, as can be seen in the previous examples, a three-input gate 2^3 (8) lines, a four-input gate 2^4 (16) lines etc.

Example 5

A logic system having three inputs is assembled as shown in Figure 8.19. Develop a truth table and describe in a sentence the relationship between the input and output. The truth table for this combination of logic gates can be drawn up as shown in Figure 8.20. Three inputs mean that the truth table must have 2^3 rows, that is eight rows. All possible combinations of input are shown in columns A, B and C. The first AND gate will give a logic 1 output only when input A and B are both logic 1. There are two such occasions as shown by column D. The second AND gate will give a logic 1 output only

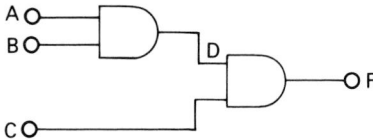

Figure 8.19 Logic network for Example 5

A	B	C	D	F
0	0	0	0	0
0	0	1	0	0
0	1	0	0	0
0	1	1	0	0
1	0	0	0	0
1	0	1	0	0
1	1	0	1	0
1	1	1	1	1

Figure 8.20 Truth table for Example 5

when input C and D are both logic 1. This occurs on only one occasion. That is, the output is at logic 1 only when all three inputs are at logic 1.

Example 6

A three input logic network is assembled as shown in Figure 8.21. Develop a suitable truth table and use this to describe the relationship between the three inputs and the output Z. The truth table for this network, which has three inputs can be constructed as shown in Figure 8.22. All possible combinations of the input are shown in columns V, W and X.

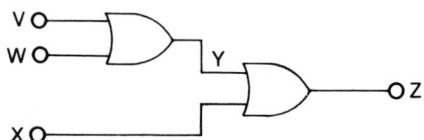

Figure 8.21 Logic network for Example 6

V	W	X	Y	Z
0	0	0	0	0
0	0	1	0	1
0	1	0	1	1
0	1	1	1	1
1	0	0	1	1
1	0	1	1	1
1	1	0	1	1
1	1	1	1	1

Figure 8.22 Truth table for Example 6

The first OR gate will give a logic 1 output when either V *or* W *or* both are at logic 1. This occurs on all but two occasions as can be seen by considering column Y of the truth table. The second OR gate will give a logic 1 output when either X *or* Y *or* both are at logic 1. This occurs on all but one occasion. Therefore, we can say that the output Z is at logic 0 *only* when all three inputs are at logic 0. If any input is at logic 1, the input Z is also at logic 1.

Logic families

The simplicity of digital electronics with its straightforward on off switching means that many logic elements can be packed together in a single

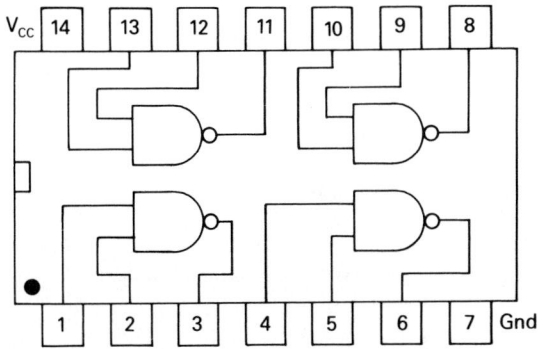

Figure 8.23 The 7400 TTL logic family

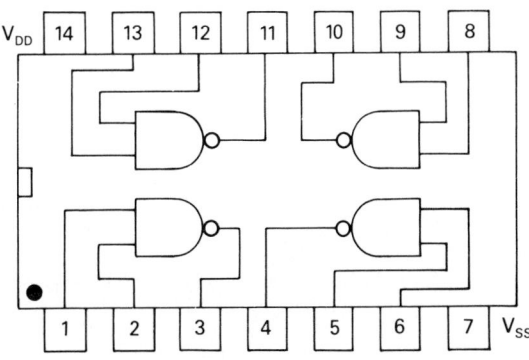

Figure 8.24 The 4011B CMOS logic family

integrated circuit and packaged as a standard dual-in-line IC as shown in Figures 6.19 and 21. Different types of semiconductor circuitry can be used to construct the logic gates. Each type is called a logic family because all members of that integrated circuit family will happily work together in a circuit.

Two main families of digital logic have emerged as the most popular with designers of general-purpose digital circuits in recent years. These are the TTL and CMOS families. The older of these is the TTL (transistor-transistor logic) family which was introduced in 1964 by Texas Instruments Ltd. The standard TTL family is designated the 7400 series. Figure 8.23 shows the internal circuitry of a TTL 7400 IC. This contains a quad 2-input NAND gate, that is, it contains four NAND gates each with two inputs and one output. Thus, with two power supply connections, the 7400 IC has 14 connections and is manufactured as the familiar 14 pin dual-in-line package. Many other combinations are available and each has its own unique number which, in this family, always begins with 74 and is followed by two other numbers. The final two numbers indicate the type of logic gate, for example, a 7432 is a quad 2-input OR gate, a 7411 a triple 3-input AND gate, as can be seen from the data sheets give in Appendix L.

The CMOS family, pronounced see-mos, is the Complementary Metal Oxide Semiconductor family of logic ICs which was introduced in 1968. The best known CMOS family is designated the 4000 series and, like its TTL equivalent, is housed in a 14 pin dual-in-line package. The 4011B is a quad 2-input NAND gate, as shown in Figure 8.24.

This is *similar* to the TTL 7400 shown in Figure 8.23, but it is not *identical* because the pin connections differ and, therefore, a TTL package cannot replace a CMOS package.

The theory of digital logic is the same for all logic families. The differences between the families are confined to the practical aspects of the circuit design. Each logic family has its own special characteristics which make it appropriate for particular applications.

Comparison of TTL and CMOS

A CMOS device dissipates about 1 mW per logic gate compared with about 20 mW for a standard TTL logic gate. Therefore, CMOS has a much lower power consumption than TTL which is particularly important when the circuitry is to be battery powered.

The output of a logic gate may be connected to the input of many other logic gates. The drive capability of a gate to hold its input at logic 0 or logic 1 while delivering current to the other gates in the circuit is called the *Fan-out* capability. The Fan-out for for TTL is ten, which means that ten other TTL logic gates can take their input from one TTL output and still switch reliably before overloading occurs. A Fan-out of fifty is typical for CMOS because they have a very high input impedance and low power consumption.

The power supply for TTL must be 5 V ± 0.25 V with a ripple of less than 5% peak to peak. A TTL device will be damaged if voltages in excess of these limits are applied. This requirement can be

Table 8.1 Properties of logic families

Property	TTL	CMOS
Power consumption	high – 20 mW	low – 1 mW
Operating current	high – mA range	low – μA range
Power supply	5 V $\pm$ 0.25 V dc	3 V to 15 V dc
Switching speeds	fast – 10 ns	slow – 100 ns
Input impedance	low	high
Fan-out	10	50

easily met by the IC fixed voltage regulators discussed in Chapter 7. CMOS devices can tolerate a much wider variation of supply voltages typically +3 V to +15 V.

Another advantage of CMOS logic circuits is that they require only about one fiftieth of the 'floor space' on a silicon chip compared with TTL. CMOS is, therefore, ideal for complex silicon chips such as those required by microprocessors and memories.

The switching times of any logic network are infinitesimal when compared with an electro-mechanical relay. However, the switching times for TTL logic are very much faster than CMOS, although both are measured in nanoseconds (10^{-9}s). The properties of each family are summarised in Table 8.1.

Working with logic

ICs of the same number will always have the same function regardless of the manufacturer and any suffix or prefix which may accompany the basic gate number. Therefore, an IC package must be replaced with another of the same number. The very high input impedance of CMOS accounts for its low power consumption but it does mean that static electricity can build up on the input pins if they come into contact with plastic, nylon or the manmade fibres of workers' clothing during circuit assembly or repair. This does not happen with TTL because the low input impedance ensures that any static charges leak harmlessly away through the junctions in the IC. Static voltages on CMOS can destroy them, and they are supplied with anti-static carriers and these should not be removed until wiring is completed. Internal protection is also provided by buffered inputs but these

cannot become effective until the supply is connected. Inputs must, therefore, be disconnected before the mains connections when disconnecting CMOS. Alternatively, the power supplies must be connected before the inputs when assembling CMOS chips. Input signals must not be applied until the power supply is connected and switched on.

When operating CMOS with normal positive logic signals VSS is the common line (OV) and VDD is the positive connection, 3 to 15 volts. Unused inputs must not be left *floating*. They must always be connected in parallel with similar used inputs, or connected to the supply rail.

Working with CMOS has created many new problems for electronic technicians. These can be overcome by

1. working on a copper plate working surface which is connected to earth,
2. ensuring that all equipment is properly earthed, and
3. wearing a conductive wrist-band which is connected to the earth of the working surface.

When these precautions are observed the problems of handling CMOS ICs can be overcome without too much difficulty.

British Standard symbols

Although the British Standards recommend symbols for logic gates, much of the manufacturers' information uses the American 'MilSpec' Standard symbols. For this reason I have reluctantly used the American standard symbols in this chapter. However, there is some pressure in the UK to adopt the B.S. symbols and for this reason the British Standard and American Standard symbols are cross-referenced in Appendix K.

Exercises

1. A voltage signal which changes smoothly and progressively between two extremes is called
   (a) a logical waveform
   (b) an analogue waveform
   (c) an interference signal
   (d) a digital waveform

2. A voltage signal which has two quite definite states, either on or off, is called
   (a) a logical waveform
   (b) an analogue waveform
   (c) an interference signal
   (d) a digital waveform

3. A single logic gate has two inputs X and Y and one output Z. The output Z will be at logic 1 only when input A and input B are at logic 1 if the gate is
   (a) a NOT gate
   (b) an AND gate
   (c) an OR gate
   (d) a NOR gate

4. Develop the truth table for an Exclusive – OR gate.

5. Develop the truth table for a NOR gate.

6. For the circuit shown in Figure 8.25 develop the truth table.

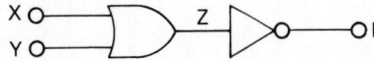

Figure 8.25 Logic network for question number 6

7. Develop the truth table for the network shown in Figure 8.26 and describe the relationship between the inputs and output.

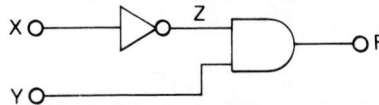

Figure 8.26 Logic network for question number 7

8. Develop the truth table for the logic system shown in Figure 8.27 and describe the relationship between the output and inputs.

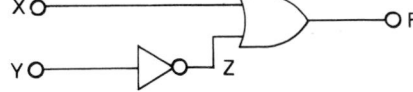

Figure 8.27 Logic network for question number 8

9. Work out the truth table for the circuit shown in Figure 8.28. Describe in a sentence the behaviour of this circuit.

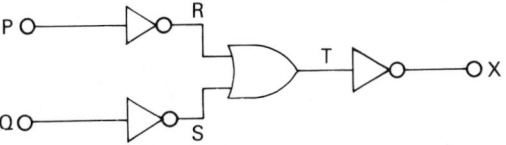

Figure 8.28 Logic network for question number 9

10. Complete the truth table for the circuit shown in Figure 8.29 and describe the circuit behaviour.

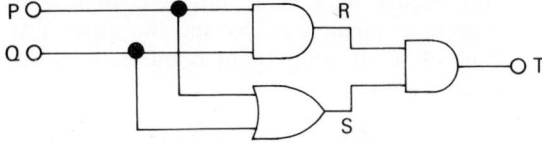

Figure 8.29 Logic network for question number 10

11. Using a truth table describe the output of the logic system shown in Figure 8.30.

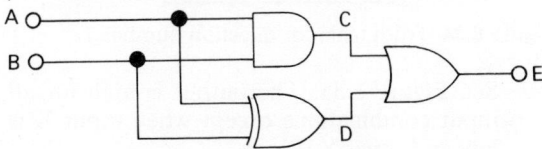

Figure 8.30 Logic network for question number 11

12. Use a truth table to describe the output of the logic network shown in Figure 8.31.

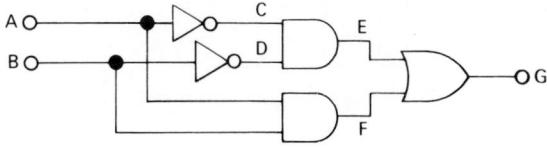

Figure 8.31 Logic network for question number 12

13. Inputs A B and C of Figure 8.32 are controlled by three separate key switches. Determine the sequence of key switch positions which will give an output at F.

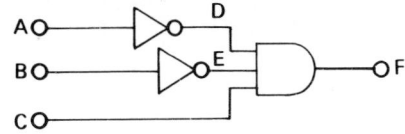

Figure 8.32 Logic network for question number 13

Solutions

1. b.
2. d.
3. b.
4. See Figure 8.4.
5. See Figure 8.6.
6. See Figure 8.33.

X	Y	Z	F
0	0	0	1
0	1	1	0
1	0	1	0
1	1	1	0

Figure 8.33 Truth table for question number 6

7. See Figure 8.34. The output is high only when the input X is low and the input Y is high. For all other input combinations the output is low.

X	Y	Z	F
0	0	1	0
0	1	1	1
1	0	0	0
1	1	0	0

Figure 8.34 Truth table for question number 7

8. See Figure 8.35. The output is high for all input combinations except when input X is low and input Y is high.

X	Y	Z	F
0	0	1	1
0	1	0	0
1	0	1	1
1	1	0	1

Figure 8.35 Truth table for question number 8

9. See Figure 8.36. The output X is at logic 1 only when input P and input Q are at logic 1. For all other input combinations the output is logic 0.

P	Q	R	S	T	X
0	0	1	1	1	0
0	1	1	0	1	0
1	0	0	1	1	0
1	1	0	0	0	1

Figure 8.36 Truth table for question number 9

10. See Figure 8.37. The output T is logic 1 only when both inputs are at logic 1. For all other input combinations the output is logic 0.

P	Q	R	S	T
0	0	0	0	0
0	1	0	1	0
1	0	0	1	0
1	1	1	1	1

Figure 8.37 Truth table for question number 10

11. See Figure 8.38. The output E is logic 1 for all input combinations except when input A and B are both logic 0.

A	B	C	D	E
0	0	0	0	0
0	1	0	1	1
1	0	0	1	1
1	1	1	0	1

Figure 8.38 Truth table for question number 11

12. See Figure 8.39. The output is high when both inputs are the same.

A	B	C	D	E	F	G
0	0	1	1	1	0	1
0	1	1	0	0	0	0
1	0	0	1	0	0	0
1	1	0	0	0	1	1

Figure 8.39 Truth table for question number 12

13. See Figure 8.40. An output is only available at F when keys A and B are off (both at logic 0) and key C is on.

A	B	C	D	E	F
0	0	0	1	1	0
0	0	1	1	1	1
0	1	0	1	0	0
0	1	1	1	0	0
1	0	0	0	1	0
1	0	1	0	1	0
1	1	0	0	0	0
1	1	1	0	0	0

Figure 8.40 Truth table for question number 13

CHAPTER 9

Transducers

A transducer is a device or element which converts one form of energy applied to the input into another form of energy at the output. The reason for wanting to do this is to convert the input signal into another form which is easier to work on, easier to amplify, easier to transmit to another place or easier to present on a display panel. It may be a device which converts mechanical vibrations into an electrical signal or which converts rotary motion into an electrical signal, as was the case with the tachogenerator shown in Figure 4.3. When used in commercial and industrial applications such as process control, the transducer generally converts a non-electrical input into an electrical output because there are many advantages to be gained by measuring non-electrical quantities by electrical and electronic instruments.

Transducers and process control

Process control and automation are concerned with handling large quantities of material whose physical and chemical properties need to be continuously monitored for quality and consistency of the finished product. Transducers convert the physical and chemical quantities into electrical signals so that the quantities can be measured. Many industrial processes create a hostile human environment and the transducer permits the quantity to be measured at a safe distance, often with much more convenience. For example, several sources of information can be observed simultaneously and compared, and the small output signals from the transducer can easily be amplified for display purposes. Extensive monitoring and control are required by industry and most systems use an appropriate transducer as the sensing element.

Measurement of strain

All materials deflect slightly when a force is applied to them, which causes tensile or compressive strain. Spanners bend slightly when a nut is tightened and bridges stretch and bend in high winds or when heavily loaded. A simple and convenient method of measuring this strain is to fix a *strain gauge* to the test material so that it experiences similar strains. In industry strain gauges are used during the development and testing of a product. For example, the behaviour of an aircraft wing in a wind tunnel can be monitored with strain gauges. For general engineering stress and strain analysis, a range of foil strain gauges such as those shown in Figure 9.1 is available.

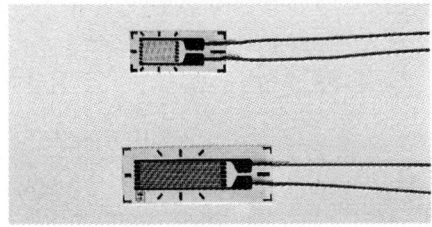

Figure 9.1 Foil strain gauges

Strain gauges

A strain gauge is a device which experiences a change of electrical resistance when it is strained (because $R = \rho l/a.\Omega$). If the gauge is stuck firmly to the surface of a much more rigid body, any changes in dimensions of the body will cause an identical but fractional change in the dimensions of the strain gauge wire.

A modern strain gauge is formed by rolling out a thin foil of the resistive material and then cutting away parts of the foil by a photo-etching process to create the required grid pattern. This method of construction has the following advantages:

1. The strain gauge is very thin.
2. The gauge has a large cross-section and, therefore, a large area of contact with the test surface.
3. Photo-etching techniques lend themselves to accurate reproduction and the production of 'matched sets' of gauges.
4. The overall dimensions of the strain gauge can be very small so that measurements of localised strain can be made.

Gauge factor

Not all strain gauges are the same. When we calculate the strain occurring in a strain gauge we must put into the calculation a factor which reflects the 'character' of the strain gauge being used. This is called the *gauge factor* and is the constant of proportionality between the applied strain and the resistance of the gauge.

$$\text{Resistance } \alpha \text{ change in dimension}$$

$$\therefore \text{ Resistance } \alpha \text{ strain}$$

$$\text{Resistance} = k \times \text{strain}$$

where k = the gauge factor (G.F.)

$$\therefore \text{Resistance} = \text{G.F.} \times \text{strain}$$

The gauge factor of most strain gauges is between 1.8 and 2.2, the total resistance varies between 60 Ω and 2000 Ω but the most common value is 120 Ω ± 0.5%

Active and passive axis

If the strain gauge shown in Figure 9.2 is strained horizontally it will cause a change in the dimensions of the majority of the gauge wire and, therefore, a change in the resistance of the wire. However, if the gauge is strained vertically the loop will simply try to open out, which will have very little influence upon the dimensions of the gauge wire and, therefore, upon the resistance. The horizontal axis of the gauge is called the *active*

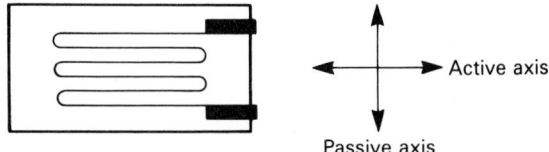

Figure 9.2 A single active axis strain gauge

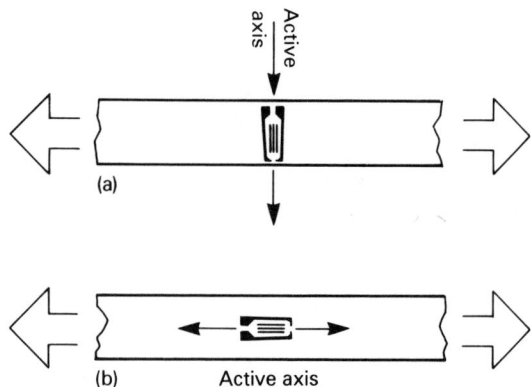

Figure 9.3 Alternative methods of mounting a strain gauge to a metal specimen

axis and the vertical axis the *passive axis*. The change of resistance will be much greater if the gauge is strained along its active axis than if the same gauge was strained along its passive axis. Figure 9.3 shows a single strain gauge mounted on to a specimen of metal which is being stretched by a tensile load as shown by the arrows. It is clear that the gauge mounted on metal (b) will produce the greatest change of resistance.

Actual measurements

A strain gauge attached to a specimen will exhibit a change in resistance when the specimen is strained. However, the change in resistance will be very small, perhaps 1 part in 10,000, and changes of this magnitude are difficult to detect directly. They are also subject to errors due to other equally small variations such as temperature changes. To overcome these problems the strain gauge is usually connected to one arm of a Wheatstone Bridge circuit as shown in Figure 9.4. The output

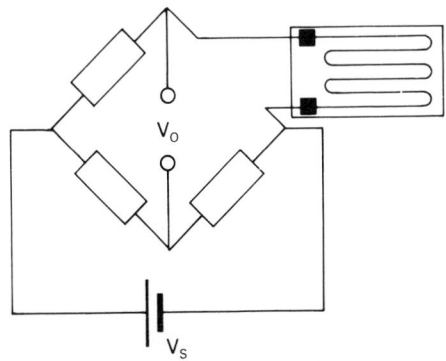

Figure 9.4 A single active gauge bridge circuit

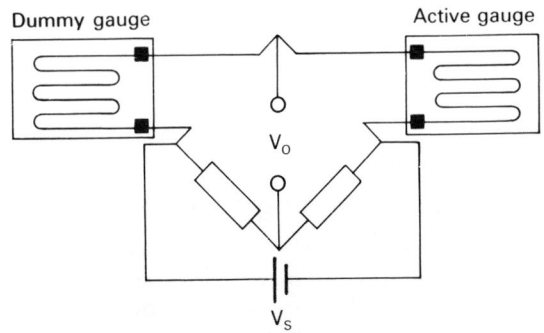

Figure 9.5 Connection of a single active gauge and dummy gauge for temperature compensation

voltage of a single active gauge bridge circuit is given by

$$V_O = \frac{V_S . \text{G.F.} e}{4} \text{ volts where}$$

V_O = the output voltage

V_S = the supply voltage

G.F. = the gauge factor

e = the strain

Example 1

A single active strain gauge is connected to a Wheatstone Bridge circuit as shown in Figure 9.4. Calculate the output voltage when G.F = 2, V_s = 20 V and the strain is 0.01.

$$V_O = \frac{V_S . \text{G.F.} e}{4} \text{ volts}$$

$$\therefore V_O = \frac{20 \text{ V} \times 2 \times 0.01}{4} = 0.1 \text{ V}$$

Temperature compensation

Changes in the temperature of the specimen to which the gauge is attached will cause changes in the resistance of the gauge which are not related to the strain. These errors in strain measurement due to temperature variations can be significantly reduced by using a second *dummy gauge* to

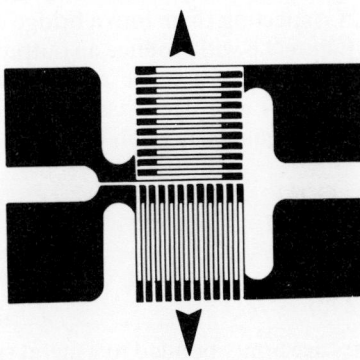

Figure 9.6 A pair of 'matched' strain gauges used for temperature compensation

compensate for them. The dummy gauge should be mounted on an unstressed specimen of the same material and connected as another arm of the Wheatstone Bridge as shown in Figure 9.5. The dummy gauge should be mounted as close as possible to the active gauge so that the temperature of both gauges is identical.

In practice there is some difficulty in finding an unstrained specimen of the same material close to the active gauge. This is overcome to a large extent by placing a dummy gauge on the same member but with its active axis at right-angles to the direction of strain. Foil strain gauges which have two strain elements at right-angles are available as shown in Figure 9.6. This ensures that the two gauges are matched for resistance and temperature and are always mounted close together. Since there is still only one active gauge, the dummy gauge being unstressed, the output voltage is still given by

$$V_O = \frac{V_S.G.F.e}{4} \text{ volts}$$

To measure bending strain

Two *active gauges* may be attached to a member as shown in Figure 9.7 to measure the bending strain of a member. If they are mounted one on either side of the member with their active axis along the length of the member, then a tensile strain will be imposed on the upper gauge and a compressive strain on the other. Both strain gauges will experience a resistance change of the same value but the upper gauge will increase while the lower one will decrease. Connecting them into a bridge circuit as shown in Figure 9.8 will produce an output voltage which is twice that of a single active gauge. The output voltage for a bridge circuit which incorporates two active gauges is given by

$$V_O = \frac{V_S.G.F.e}{2} \text{ volts}$$

Example 2

Two gauges are firmly bonded to a metal specimen to measure bending strain as shown in Figure 9.7. Both gauges have their active axis along the length of the specimen. The input voltage is 20 V, the gauge factor 2 and the strain 0.01. Calculate the output voltage

$$V_O = \frac{V_S.G.F.e}{2} \text{ volts}$$

$$\therefore V_O = \frac{20 \times 2 \times 0.01}{2} = 0.2 \text{ V}$$

It can be seen that the output is twice that of Example 1 which had only one active gauge. However, temperature compensation will not occur with this arrangement unless the resistors in Figure 9.8 are replaced with dummy gauges.

To measure tensile strain

Two active gauges can be used in a bridge circuit to measure the tensile strain in a member under

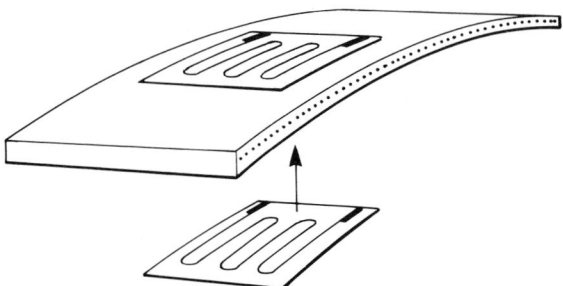

Figure 9.7 Two active gauges attached to a member measuring bending strain

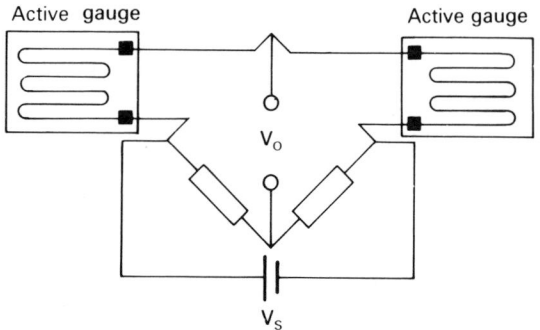

Figure 9.8 Two active gauges connected to a bridge circuit for measuring bending strain

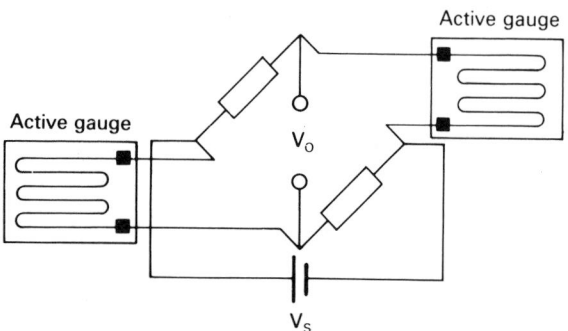

Figure 9.9 Two active gauges connected to a bridge circuit for measuring tensile strain

tension. The gauges should be attached to the specimen as before and shown in Figure 9.7, and connected into a bridge circuit as shown in Figure 9.9. In this case the gauges are connected into opposite arms of the bridge and the output is once more given by the formulae $V_O = V_S.G.F.e/2$ volts because these are two active gauges.

Measuring the output voltage

In practice the output voltage from the bridge circuit is pretty small, which means that we can either use a sensitive voltmeter or amplify the bridge output in order to use a less sensitive voltmeter. The second option is most often preferred because it is cheaper. The 741 operational amplifier which we have already considered in Chapter 7 makes an excellent strain gauge amplifier. This is because the output voltage from the bridge occurs as the voltage difference between two points which can be connected to the inverting and non-inverting input of the op amp as shown in Figure 9.10.

Example 3

Two active gauges are bonded to a metal specimen held in tensile strain and connected to the circuit shown in Figure 9.10. The gauge factor is 2 and the gain Av of the op amp is 100. Calculate the tensile strain on the specimen when the voltmeter is reading 2 V.

Because there are two active gauges the appropriate formula is

$$V_O = \frac{V_S.G.F.e}{2} \text{ volt}$$

transposing this formula for strain we have

$$e = \frac{2.V_O}{V_S.G.F.} \text{ (strain has no units)}$$

V_O is the output from the bridge which is the input to the op amp. The op amp gain is 100 and, therefore, the input is 100 times less than 2 V i.e. 0.02 V.

$$\therefore V_O = 0.02 \text{ V}$$

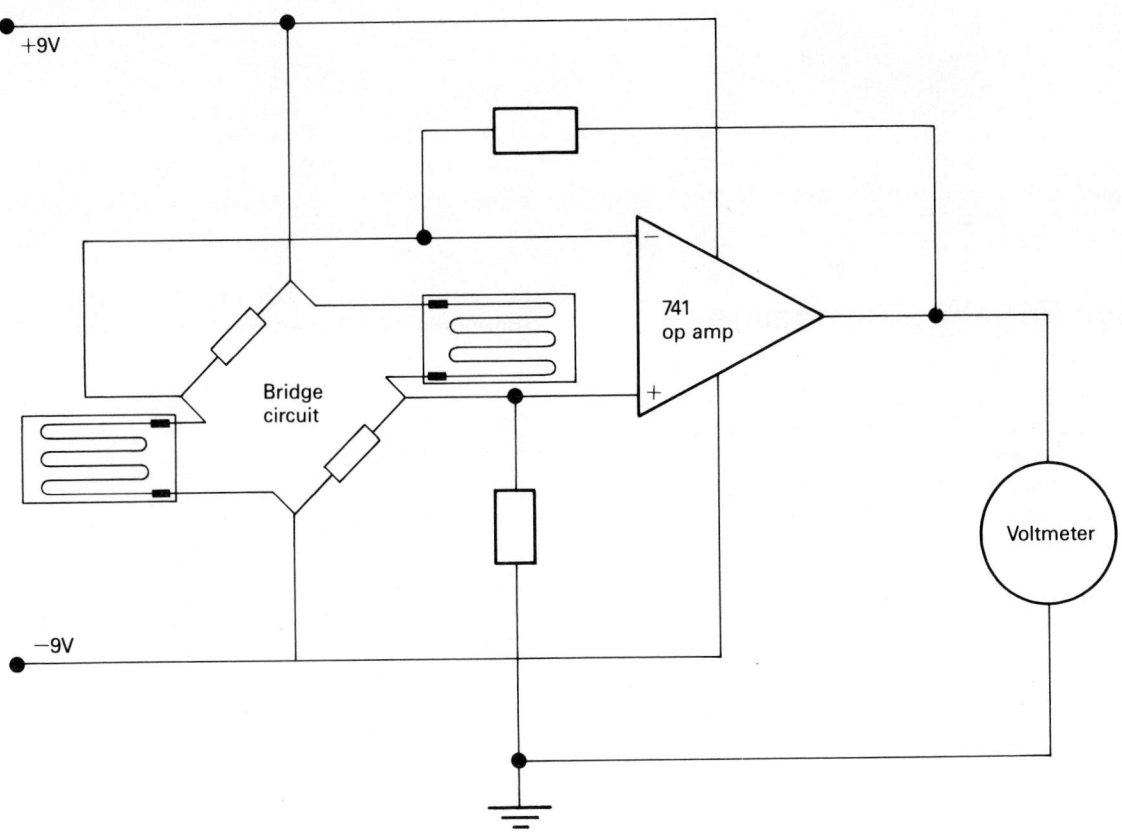

Figure 9.10 Using an OpAmp as a strain gauge amplifier

V_s is the supply to the bridge which in this case is the voltage difference between +9 V and −9 V i.e. 18 V.

$$\therefore V_S = 18 \text{ V}$$

Substituting these values into the formula we have

$$e = \frac{2 \times 0.02 \text{ V}}{18 \text{ V} \times 2} = 0.001$$

Therefore the tensile strain on this specimen is 0.001.

Strain gauge amplifiers such as that shown in Figure 9.11 are available commercially. The one shown encapsulates the amplifier in a 24 pin DIL package and can be used to interface any transducer with a resistive bridge configuration.

W. 20
L. 33
D. 13

Figure 9.11 A commercially available strain gauge amplifier

Bonding the strain gauge

Probably the weakest part of a strain gauge measurement system is the adhesion of the gauge to the metal under test. The electrical connections can be tested and remade if necessary but there is no way to determine if a strain gauge is only partly bonded to the metal and giving false readings. Therefore, bonding procedures must be scrupulously followed if the strains experienced by the metal under test are to be transmitted accurately to the gauge.

1. Specimen surface preparation

An area larger than the strain gauge should be cleared of all paint, rust etc., and finally smoothed with a fine-grade emery paper or fine sand blasting to provide a sound bonding surface. The area should be degreased with a solvent and finally neutralised with a weak detergent solution. Tissues or lint-free cloth should be used for this operation, wetting the surface and wiping off with clean tissues or cloth until the final tissue used is stain free. Care must be taken not to wipe grease from a surrounding area on to the prepared area or to touch the surface with the fingers. This final cleaning should take place immediately prior to the installation of the gauge.

2. Strain gauge preparation

By sticking a short length of sellotape over the strain gauge, it can be lifted from a flat surface, taking care not to bend the gauge sharply. Holding both ends of the tape, orientate the gauge on the

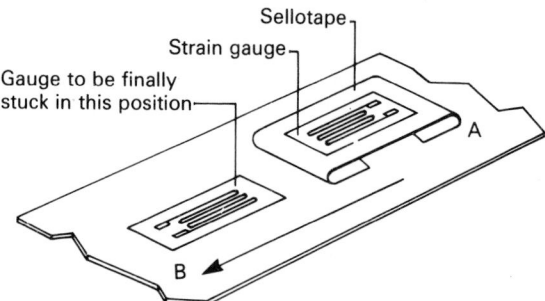

Sellotape
Strain gauge
Gauge to be finally stuck in this position
A
B

Figure 9.12 Bonding a strain gauge to the metal specimen

prepared site and stick down the end of the sellotape furthest from the connection tags. Bend the other end of the tape back upon itself to expose the underside of the gauge as shown in Figure 9.12.

3. Adhesives

Two alternative types of adhesive are available, quick-set epoxy and cyanoacrylate. When using epoxy adhesive apply a smooth thin coat to the whole underside of the strain gauge. Unstick end A of the sellotape, roll the gauge over in the direction of arrow B and press it down firmly into position, wiping the excess adhesive to the outside edges. Care should be taken to leave an even layer of adhesive with no air bubbles left under the strain gauge. Cover the gauge and apply a light weight or clamp until the adhesive sets. Remove the sellotape by slowly and very carefully pulling it back over itself, starting at the end furthest from the connection tags. Do not pull upwards.

If cyanoacrylate adhesive is used, stick one end of the tape down to the specimen, completely up to the gauge. Drop a small amount of adhesive in the 'hinge' point formed by the gauge and the specimen. Starting at the fixed end, with one finger push the gauge down, at the same time pushing the adhesive along the gauge in a single wiping motion until the whole gauge is stuck down. Apply pressure with one finger over the whole length of the gauge for approximately one minute. Leave for a further three minutes before removing the sellotape.

Cyanoacrylate adhesives must be handled with care because they can also bond fingers or eyelids together. Accidental contact with the adhesive should be washed away with water immediately. Accidentally bonded fingers can be peeled apart by using soap, hot water and a blunt parting tool such as the handle of a spoon.

4. Wiring

The lead out wires should be soldered and insulated with heat shrink sleeve or similar. The lead out wires are fragile and should be handled with care.

5. Protection

After bonding the strain gauge and making the electrical connections, the gauge can be protected against humidity and moisture with a coat of air-drying varnish. For a more permanent installation, the gauge and its electrical connections can be encapsulated by spreading a layer of silicone rubber compound over them with a spatula, so that they become embedded in a flexible coating which dries in about 24 hours.

Measurement of pressure

Pressure is a measure of the force exerted per unit of cross-sectional area. Pressure can, therefore, be measured by monitoring the force exerted on a thin steel diaphragm which flexes under pressure as shown in Figure 9.13. The displacement X of the diaphragm is proportional to the pressure. If a strain gauge is attached to the diaphragm a variety of indicating and recording instruments can be

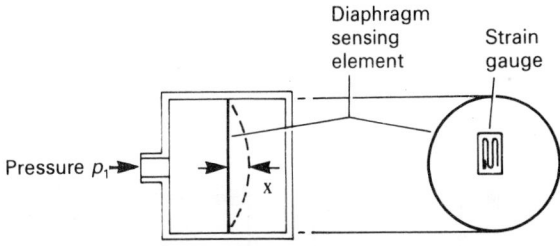

Figure 9.13 Construction of a pressure transducer

driven so that the pressure can be measured at some distance from the pressure transducer.

Using stainless steel as a diaphragm material results in a rugged pressure gauge able to withstand accidental overload. This type of construction is easy to waterproof and can be beam welded to withstand a variety of corrosive media. However, they are not available in diameters of less than 3.2 mm but small diameters can be achieved if the actual diaphragm is made from a silicon chip which incorporates a strain gauge. This technique permits pressure gauge diaphragms to be manufactured which have active diameters as small as 0.75 mm. The major disadvantage with silicon diaphragms is the difficulty in providing a seal and the tendency of the brittle silicon crystal to crack or shatter if overloaded.

Piezoelectric pressure transducers

An alternative to using a strain gauge as the pressure-sensing element is to use a piezoelectric crystal. If a cube of quartz crystal is squeezed across two faces, an electrical charge proportional to the pressure is generated across the faces at right-angles to the pressure as shown in Figure 9.14. The voltage is generated directly as a result of the squeezing effect of the applied pressure. Quartz crystals are, therefore, called piezoelectric crystals because the Greek word *piezein* means to squeeze. The piezoelectric effect was discovered in 1880 by the brothers Pierre and Jaques Curie (Pierre was the husband of Marie Curie, the Nobel Prize winner). They also discovered that the reverse effect is true, that is, a dimensional change takes place when an electrical potential is applied to a quartz crystal.

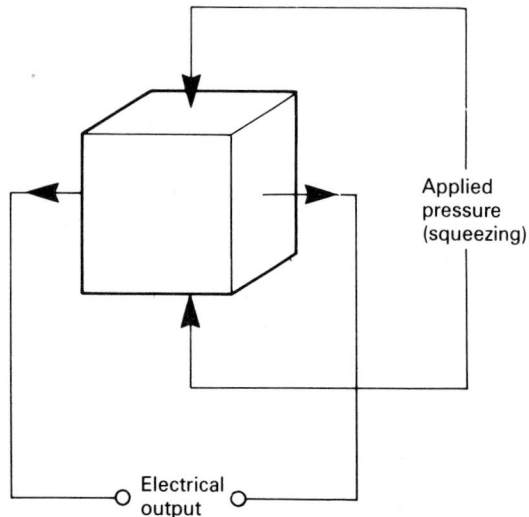

Figure 9.14 Piezoelectric effect of a quartz crystal

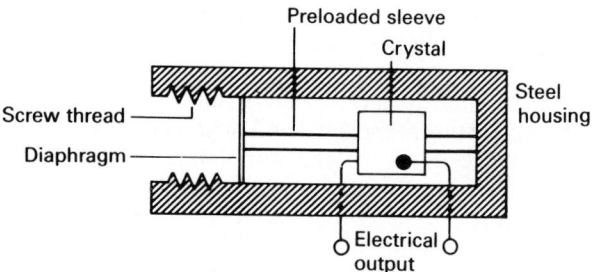

Figure 9.15 Construction of a piezoelectric pressure transducer

In the piezoelectric pressure transducer, the crystal is contained in a steel housing and held in place by a preloaded sleeve, which is a thin metal cylinder as shown in Figure 9.15. The pressure is applied to a thin steel diaphragm which deflects and converts the pressure into a compressive force on the quartz crystal. This produces a voltage at the output terminals which is converted into a usable output voltage signal by an operational amplifier, as shown in Figure 9.16.

A piezoelectric transducer will only respond to *changes* in applied pressure and, therefore, this transducer is only suitable for dynamic pressure measurements. There is no output when a steady pressure is applied.

Bourdon tube pressure transducer

The principle of the Bourdon tube pressure gauge as a test instrument has already been described in Chapter 5. The Bourdon tube pressure gauge is a

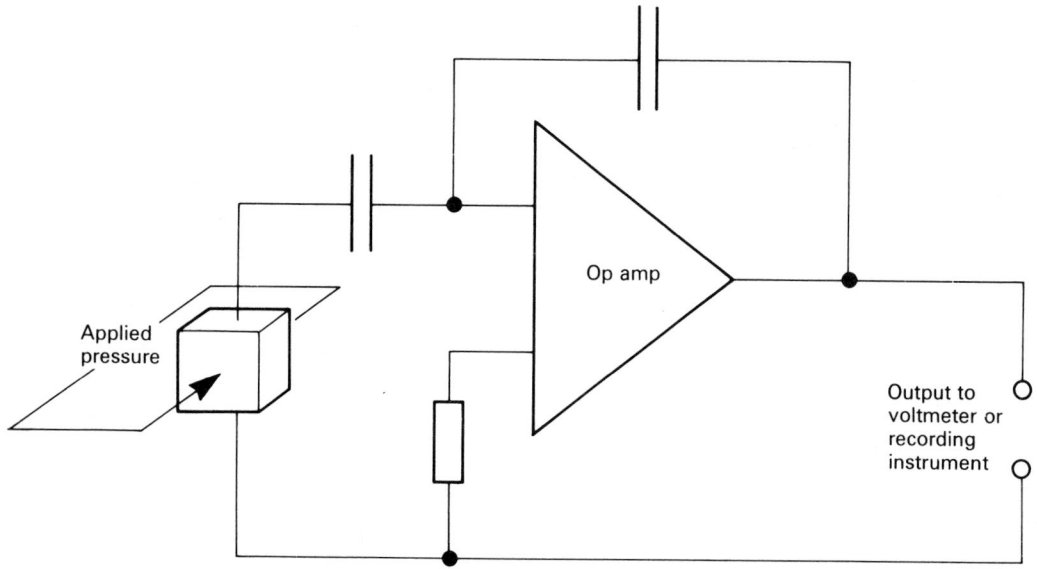

Figure 9.16 OpAmp used with a piezoelectric transducer

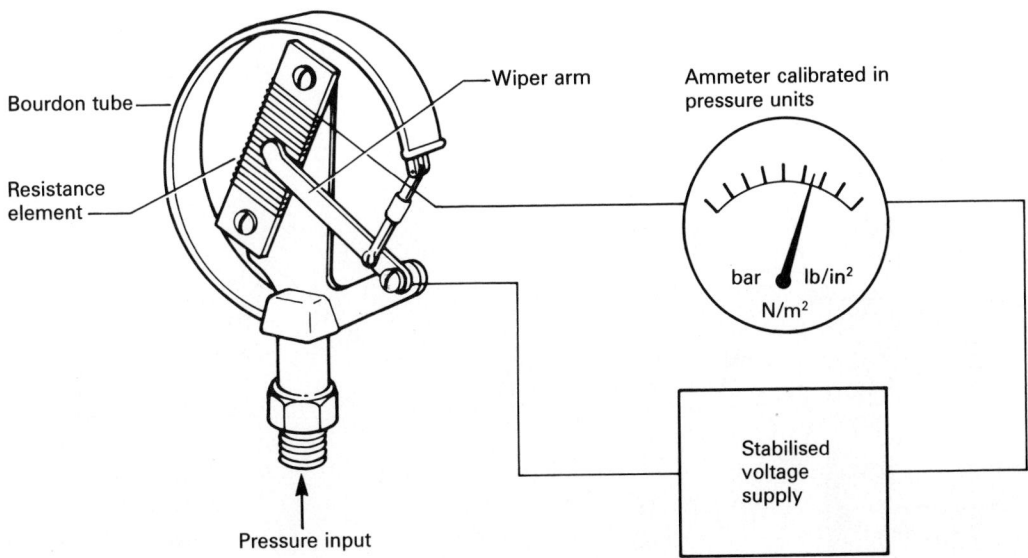

Figure 9.17 Construction of a Bourdon tube pressure transducer with an electrical output

rugged, reliable, trouble-free instrument in almost universal use. The only disadvantage is that because it is entirely mechanical, it cannot be used for remote reading and it cannot produce a signal which can be used in an automatic pressure control system. However, the transducer itself, the Bourdon tube, need not necessarily be followed by mechanical signal conditioning as is the case with the Bourdon tube pressure gauge. The output from the transducer can be converted into an electrical signal if the end of the Bourdon tube is linked to the wiper arm of a variable resistor as shown in Figure 9.17. When pressure is admitted to the Bourdon tube, the tube straightens out and causes the wiper arm to move over the resistance element. This varies the resistance in the electrical circuit which, if it is supplied by a stabilised voltage supply, will vary the current in the circuit. The ammeter scale can be calibrated in pressure units for direct reading of pressure at some distance from the transducer.

Microphone pressure transducer

Sound is transmitted by pressure waves in air and, therefore, a microphone can be used as a noise-measuring transducer since it will convert a pressure wave into an electrical signal.

When a pressure wave strikes the diaphragm of the moving coil microphone shown in Figure 9.18, the diaphragm is deflected slightly. A delicate coil wound on to the centre of the diaphragm is also deflected in the magnetic field of the permanent magnet. When any coil moves in a magnetic field, a voltage is induced in the coil. As the diaphragm vibrates in and out in response to the pressure wave fronts, the coil moves in the magnetic field and, therefore, a voltage is induced in the coil which is proportional to the sound.

A piezoelectric microphone uses a crystal of piezoelectric material in place of the moving coil. The diaphragm squeezes the mechanical axis of the crystal which produces an electrical output along the electrical axis.

A *loudspeaker* is a transducer which converts an electrical signal into a pressure wave. The construction of a loudspeaker is essentially the same as a microphone except that the diaphragm is larger. Passing a current through the moving coil of Figure 9.18 will produce a magnetic field which will react with the permanent magnetic field causing the diaphragm to vibrate. The vibrating diaphragm produces pressure wave fronts which the human ear can identify as sound.

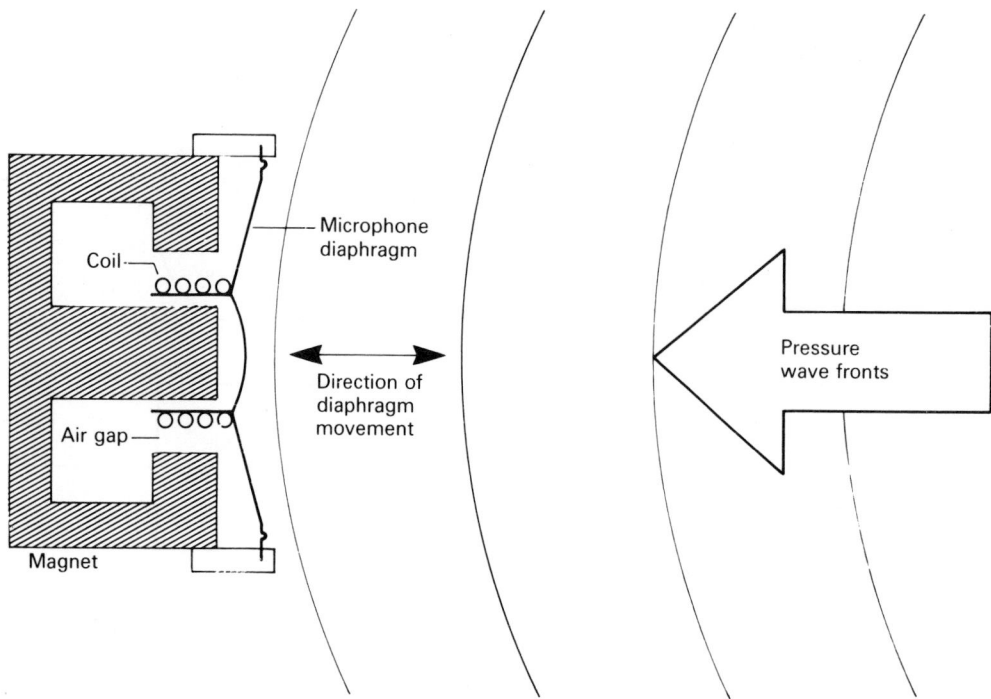

Figure 9.18 Moving coil microphone pressure transducer

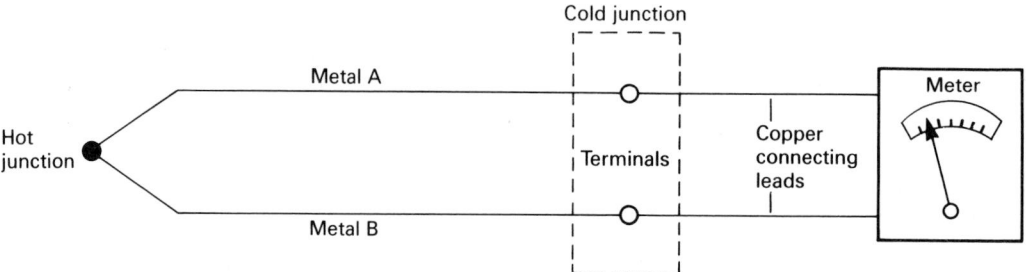

Figure 9.19 A simple thermocouple circuit

Measurement of temperature

Heat is a form of energy because it is capable of doing work. Temperature is a measure of the degree of hotness of an object. Temperature can be measured and its hotness indicated by a thermometer, but a simple thermometer cannot provide an indication of the temperature at a position which is remote from the sample point. For a remote indication or for the temperature measurement to form part of an automatic temperature control system we need a temperature transducer to convert the heat energy into an electrical signal.

Thermocouple temperature transducer

A thermocouple temperature transducer consists of a single pair of insulated wires of dissimilar metal fused together at one end called the hot junction, as shown in Figure 9.19. The wire ends which are not joined are called the cold junction and are maintained at room temperature. A difference of temperature between the hot and cold junctions produces a difference of potential and, therefore, a current in milliamperes flows.

Table 9.1   Thermocouple types and temperature ranges

Thermocouple	*Material	Max. temperature range (°C)
Base-metal		
type T	Copper–constantan	400
type E	Chromel–constantan	1000
type J	Iron–constantan	1000
type K	Chromel–alumel	1300
type N	Nicrosil/nisil	1300
Rare-metal		
type S	Platinum–platinum/10% rhodium	1600
type R	Platinum–platinum/13% rhodium	1600
type B	Platinum/30% rhodium–platinum/6% rhodium	1800

*constantan = copper/nickel; chromel = nickel/chromium;
alumel = nickel/aluminium;nicrosil = nickel/chromium/silicon;
nisil = nickel/silicon.

Variations of temperature produce approximately proportional variations in voltage which can be used to give an indication of the temperature. This effect was discovered in 1821 by Thomas Seebeck, a German physicist.

To identify the various wire combinations, the thermocouples are identified by the letters of the alphabet as shown in Table 9.1. Each combination of metals generates a different voltage and has a different maximum operating temperature as shown by Table 9.1. Base metal thermocouples in general generate higher voltages for the same temperature difference, but rare metal thermocouples can withstand higher maximum temperatures. For example, the maximum operating temperature of a type T thermocouple is 400°C and at this temperature it will generate a voltage of about 20 V. A type R thermocouple has a maximum operating temperature of 1600°C and at this higher temperature it generates a lower voltage, about 10 V.

However, the voltage generated by any thermocouple is very respectable and is certainly capable of easy detection and display or for operating an electromagnetic valve in a flame failure device. The actual generated voltage of different types of thermocouple at all possible temperatures are given in the BS.4937.

It has become the modern practice to construct the thermocouple in a similar way to an MICC cable. The conductors are insulated with mineral insulation and are contained in a stainless steel tube varying in size from 0.25 mm to 3.0 mm

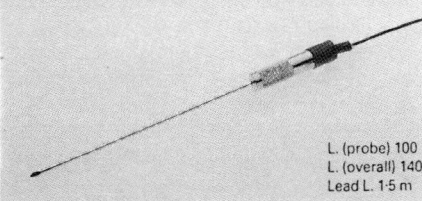

L. (probe) 100
L. (overall) 140
Lead L. 1·5 m

Figure 9.20   Type T hypodermic thermocouple probe

diameter. Figure 9.20 shows a type T thermocouple probe with a stainless steel probe of 1.6 mm diameter and 100 mm length. This thermocouple generates approximately 42 μV/°C in the temperature range 0°C to 250°C. Therefore, over a temperature rise of 200°C, this thermocouple will generate a voltage of 200°C $\times$ 42 μV/°C = 8·4 mV.

Thermistor temperature transducer

A thermistor exhibits a very large resistance change with a change in temperature. Although some thermistors increase in resistance in response to an increase in temperature (PTC), most thermistors exhibit an inverse relationship between temperature and resistance (NTC). That is, an increase in temperature causes a decrease in the resistance of the thermistor as shown by Figure

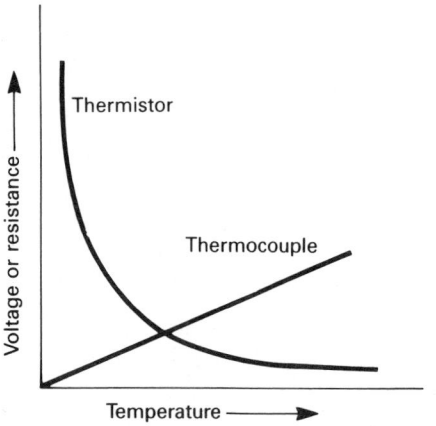

Figure 9.21 Graph showing the non-linearity of the response of a thermistor to changes in temperature

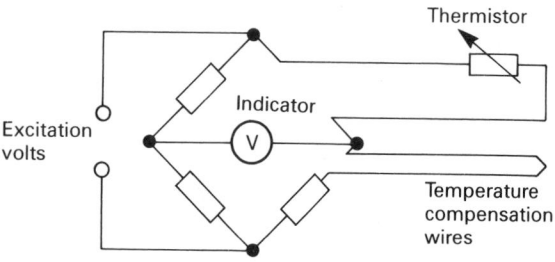

Figure 9.22 A simple thermistor circuit

9.21. Since the thermistor does not generate a voltage like the thermocouple, but changes resistance in response to changes in temperature, it is normally used as one arm of the Wheatstone Bridge as shown in Figure 9.22. As the thermistor resistance changes in response to the change in temperature the bridge becomes unbalanced and a voltage appears across the indicator.

The major advantage of the thermistor over the thermocouple is its *dramatic* change in resistance to a small change in temperature. This makes it a very *sensitive* temperature-sensing device. One major disadvantage of the thermistor is the non-linearity of the resistance change to temperature, but recently compensation networks have become available to linearise the thermistor output. It is also possible to connect the output voltage of the thermistor bridge to an analogue-to-digital converter and to use the digital output as the input to a computer. A program can then be written to linearise the thermistor output and present the temperature measurement on a suitable display.

PTC thermistors are usually used for temperature compensation and overheating protection while NTC thermistors are most often used for temperature measurement. They are manufactured from oxides of copper, manganese, nickel, cobalt and lithium, blended to give the required temperature to resistance characteristic and presented as beads, wafers or rods as shown in Figure 9.23. They can be manufactured as extremely small devices and are, therefore, ideally suited to monitoring temperature in the most inaccessible places such as the windings of an electric motor or the core temperature of power supply cables, as well as providing current-limiting protection for electronic circuits.

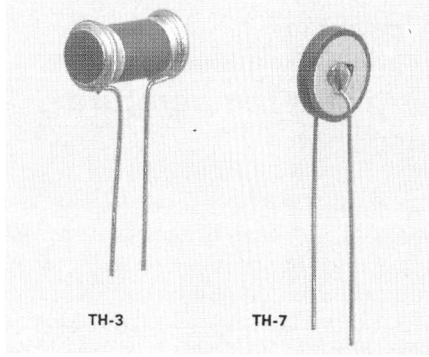

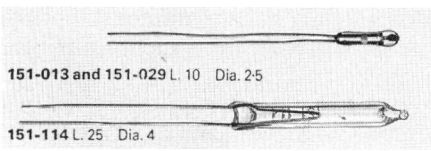

Figure 9.23 A selection of thermistors

Measurement of liquid level

The measurement of liquid level in a container is an important measurement in industry and is often used in the chemical mixing process. The simplest and most frequently used level sensor is the simple float switch. A float, moving in response to the liquid's surface, operates a switch to indicate that the liquid has reached a predetermined level. The float of a modern float switch contains a small magnet which acts upon a magnetic reed switch. Figure 9.24 shows a typical float switch. The float, which looks like a cotton bobbin, rises up the central cylinder in response to the liquid level and activates a reed switch contained in the central cylinder. The reed switch is hermetically sealed and can, therefore, operate millions of times without failure because the contacts are not exposed to atmospheric dust and corrosive fumes. The float switch shown is designed for mounting in the top of a container, but side- and bottom-mounted switches are also available. They are easily installed, reliable and require virtually no maintenance, which is an important consideration in industry.

When it is required to monitor a changing liquid level, the float can be made to slide up and down a long central stem containing many reed switches. The level can then be determined by observing the number of operated switches. For more exact readings of liquid level the float may be attached to an arm which operates a variable resistor, as

Figure 9.24 A float switch

shown in Figure 9.25. The changing resistance indicates the liquid level. This type of sensor is used to measure the fuel level in the petrol tank of a motor car.

Measurement of fluid flow

Liquid flow is relatively easy to measure, and probably the most common method used by industry is the differential pressure flowmeter or head

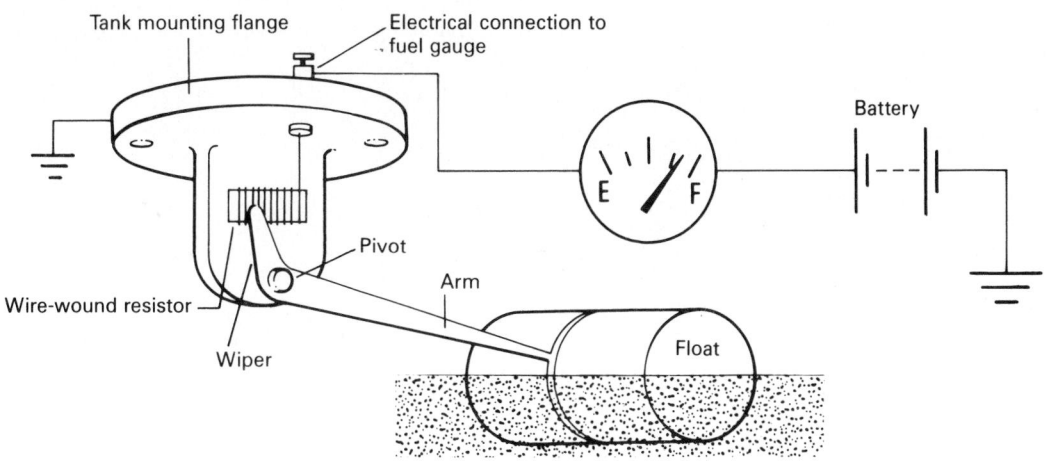

Figure 9.25 Fuel gauge float switch

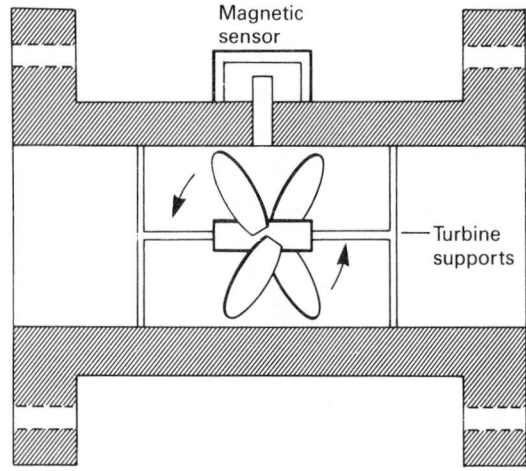

Figure 9.26 A turbine flowmeter

L. overall = 89 mm
H. overall = 51 mm
W. overall = 37 mm

Figure 9.27 A liquid flow meter

meter. When liquid passes over a curved surface or through a smaller opening than the pipe which carries it, the velocity of the liquid increases and the pressure decreases proportionately. Measuring the difference in pressure before and after the change gives an indication of the rate of flow. This principle was discovered by the Swiss mathematician Daniel Bernoulli in 1738. Knowing the rate of flow, and the cross-section of the pipe, we can calculate the flow of the liquid in gallons per minute or some other convenient unit.

Velocity flowmeters measure the rate of flow directly. The most common of these is the turbine flowmeter shown in Figure 9.26. The liquid flow causes the turbine to rotate at a speed which is proportional to the fluid flow. Rotational speed can be sensed by a magnetic proximity switch such as that shown in Figure 5.11, and the output displayed in suitable units.

The spinning turbine in the flowmeter shown in Figure 9.27 breaks a photoelectric beam, giving a pulsed output which is proportional to the rate of flow. The top chamber contains an LED and a logic chip on to which is integrated all the circuitry necessary for detection, amplification and pulse shaping. The turbine motor is in direct contact with the fluid and, therefore, this type of flowmeter can only be used with clean fluids.

When mounting flowmeters, a straight length of pipe must be available on the input side of the meter to prevent problems caused by fluctuating pressures as fluids negotiate curves. The length required varies between four and fifty times the pipe diameter depending upon the type of flowmeter. A short length of straight pipe is also recommended on the discharge side.

CHAPTER 10

Communication and security systems

Simple communication systems

The purpose of any communication system is to convey information between two physically remote points. Any communication system must include a transmitter for sending the information, a transmission circuit or medium and a receiver. To transmit signals comprehensible to the human ear the system must produce an audible note. The simplest system is a morse code or buzzer which produces a sound whenever a key activates the circuit. This sound may be of one frequency only but for the transmission of speech a complex mixture of frequencies is required and, therefore, the communication system must be capable of accommodating a wide range of frequencies. The range of frequencies required by the communication system is called the bandwidth: the simpler the system, the smaller the bandwidth.

The bandwidth of speech is from 30 Hz to 5 kHz. Music requires an even greater bandwidth from 20 Hz to 20 kHz. The telephone system is designed to transmit on a limited bandwidth from 300 Hz to 3.4 kHz. This is suitable for acceptable voice reproduction but, because the upper and lower frequencies are cut off, the quality of reproduction is not perfect.

Simple telephone circuit

A telephone handset contains a receiver, the ear piece and a transmitter for speaking into. The transmitter converts the pressure waves from the spoken word into analogue electrical signals, which are transmitted along conductors to the receiver of another handset which converts the electrical signals into sound waves. This provides the basis of a simple two-way communication system and a suitable circuit is shown in Figure 10.1.

The modern telephone system

One of the limitations of the simple telephone circuit is that a separate cable is required for each telephone conversation. If there are more people in London wanting to call someone in Edinburgh than there are lines between London and Edinburgh, then someone will have to wait. This problem can be resolved by putting up more lines, as was the case before 1940 when main roads and railways were lined with telegraph poles carrying dozens of wires, or alternatively, more than one conversation can be sent down each line. This is the modern method which is called *multiplexing*. Each conversation is put on to a carrier wave of a different frequency and all these different frequencies are sent down the line at the same time without

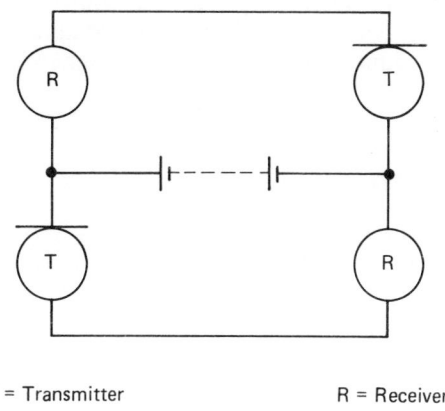

T = Transmitter   R = Receiver

Figure 10.1 A simple telephone circuit

interfering with each other. At the receiving end a filter is used to pick out a particular frequency and, therefore, a particular conversation. Another receiver can be tuned to a different frequency and, therefore, select a different conversation.

Optical fibres

For a long time telephone lines have used copper cables, but these will eventually be replaced by optical fibres which can carry more conversations with reduced line losses. Line loss is called attennation.

Glass fibre cables are about 0.1 mm in diameter and trap the light signals inside the very pure optical-quality glass fibres, using total internal reflections. The laser light signals bounce their way along the cable at the speed of light, always being reflected when they strike the inside of the cable wall in the same way that light is reflected in the prisms of binoculars. The optical-quality glass fibres are so pure that a 2 km length of cable will absorb less light than a sheet of window glass.

Some attennation (line loss) does occur and eventually the signal strength will decrease. To prevent this, amplifiers called repeaters or boosters are installed along the route, about every 30 km of line length, to restore the signal strength. Copper cables require repeater stations much closer together. The repeaters, while restoring the required signal strength, unfortunately also amplify the unwanted random noise in the system which sounds like a hiss or crackle in the receiver. In some situations, such as on long lines, the sound quality can deteriorate until the conversation is incomprehensible. This random, unwanted noise can be eliminated by using digital signals as was discussed in Chapter 8.

System X

This is the name given to a new British Telecom telephone system using digital electronic exchanges which was introduced in 1978 and will eventually replace all of the existing exchanges. All signals will be handled by an *integrated digital network* computer system. Speech signals will be transmitted in digital form. The analogue voltages generated by voice patterns will be converted to digital form by analogue-to-digital converters. All signals entering the exchange, voice, computer and fax machine data, telex, Datel and Prestel will be handled by the same computer-type circuits. The development of this system will mean that the exchange of all types of information over the telecommunications network will be faster and more simple.

Microwave telephone links

Microwaves are radio waves with a large capacity to carry information. They have a very small wavelength and can, therefore, be easily focused into a narrow beam by a dish aerial of about 2 to 3 metre diameter. Almost half the trunk calls in the UK are transmitted by terrestrial microwave links which cross the country in line-of-sight, hops between dish aerials mounted on tall towers built on hilltops. The received signal is boosted or amplified before being transmitted on to the next hop. The British Telecom tower in London is the nerve centre of this system.

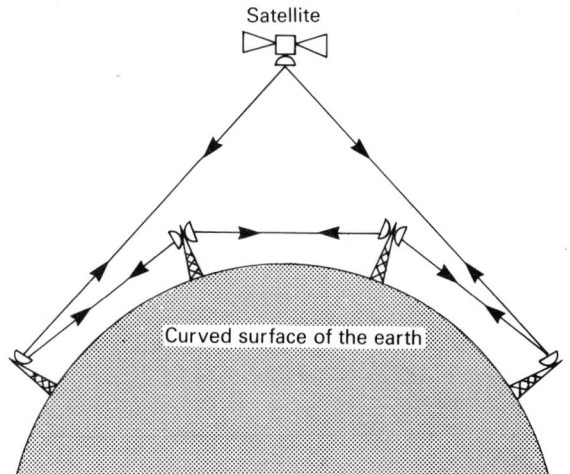

Figure 10.2 Microwave links between line of sight aerials and between earth stations linked by satellite

Satellite telephone links

Satellites can be used to transmit signals between subscribers on different continents. Two thirds of all intercontinental telephone calls are now handled by the satellite network. A typical intercontinental call goes from the caller to a local exchange where it is routed to an international exchange and then on to an earth station. The earth station transmits the signal by microwaves up to the satellite where it is retransmitted to an earth station in another country. The call is then routed via international and local exchanges to make the link between the called number and caller.

Despite the expense involved in building the satellite and placing it in orbit, it is usually cheaper than a network of transmitters and receivers which can 'see' each other over the curved surface of the earth.

Satellite communications are also discussed later in this chapter under the sub-heading Geostationary Satellite Communications.

Mobile telephones

This recently developed system is called cellular radio because the country is divided into small areas or cells. Each cell has its own low-powered radio transmitter which is linked through regional exchanges to the existing telephone system. When a caller travels from one cell to another, a computer system switches the signal to the next cell so that the call can continue uninterrupted.

Telephone at home

The installation of telecommunication equipment could, for many years, only be undertaken by British Telecom engineers. However, this monopoly was relaxed by H.M. Government from January 1985, which created potentially new markets for the retailer to supply, and the electrical contractor to install, telecommunication equipment.

On new installations the electrical contractor or competent installer may install sockets and the associated wiring to the point of intended line entry, but the connection of the incoming line to the installed master socket must only be made by a B.T. engineer.

On existing installations, additional secondary sockets may be installed to provide an extended plug-in facility. Any number of secondary sockets may be connected in parallel but the number of telephones which may be connected at any one time is restricted.

Each telephone or extension bell is marked with

Table 10.1   Telephone cable identification

code	base colour	stripe
G.W	green	white
B.W	blue	white
O.W	orange	white
W.O	white	orange
W.B	white	blue
W.G	white	green

Table 10.2   Telephone socket terminal identification

Terminals 1 and 6 are frequently unused and therefore 4 core cable may normally be installed.
Terminal 4 on the incoming exchange line is only used on a PBX line for earth recall.

Socket terminal	circuit
1	spare
2	speech circuit
3	bell circuit
4	earth recall
5	speech circuit
6	spare

a Ringing Equivalence Number (REN) on the underside. Each exchange line has a maximum capacity of REN 4 and, therefore, the total REN of all the connected telephones must not exceed 4 if they are to work correctly. If REN 4 is exceeded the volume of one or all of the ringing devices will be reduced or they may not work at all.

An extension bell may be connected to the installation by connecting the two bell wires to terminal numbers 3 and 5 of a telephone socket. The extension bell must be of the high impedance type having a REN rating. All equipment connected to a B.T. exchange line must display the green circle of approval.

The multicore cable used for wiring extension socket outlets should be of a type intended for use with telephone circuits, which will normally be between 0.4 mm and 0.68 mm in cross-section. A four-core cable should be run from the master socket outlets to each subsequent secondary socket outlet in the form of a radial circuit, and connected as shown in Figure 10.3.

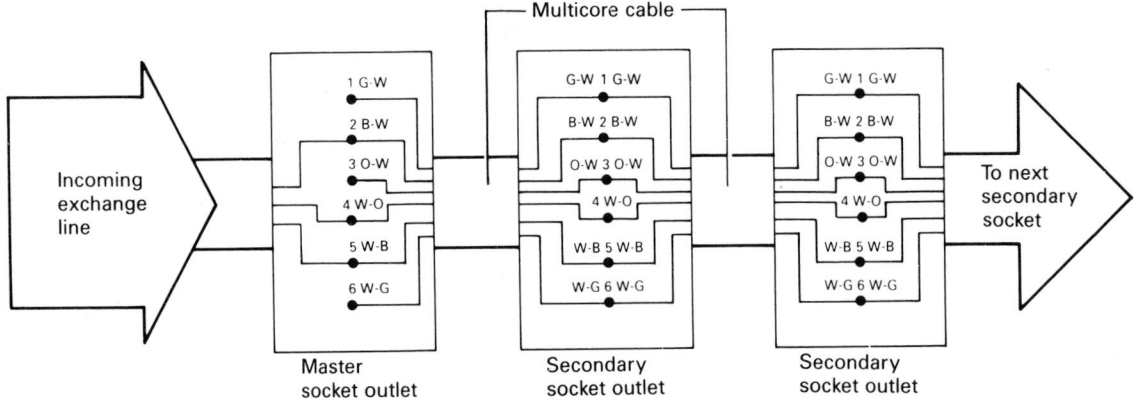

Figure 10.3 Telephone socket outlet connection diagram

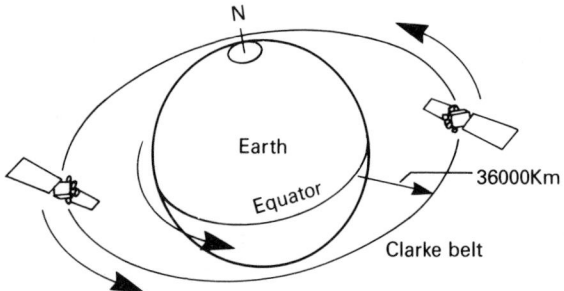

Figure 10.4 Satellites in geostationary orbit

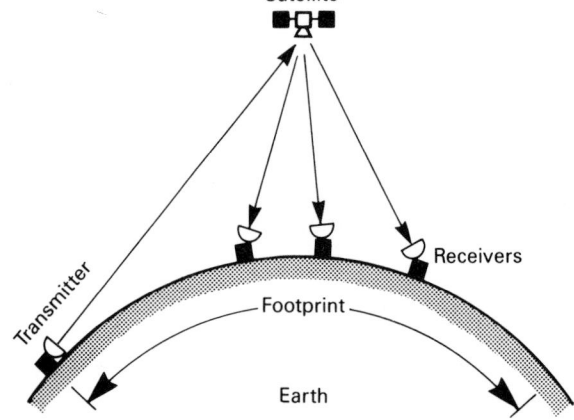

Figure 10.5 Using satellites to communicate around the curved surface of the earth

Geostationary satellite communications

A satellite placed in a circular orbit around the earth at a distance of 36,000 km (22,300 miles) above the equator will appear stationary to an observer on earth if the orbital speed of the satellite is 'matched' to the earth's surface speed.

These are the conditions for an earth stationary (geostationary) orbit described in 1945 by Arthur C. Clarke, the British writer and scientist.

In this orbit the satellite will always remain over the same point on the surface of the earth. If the satellite is equipped with a radio transmitter it can *illuminate* a large area of the earth with radio waves which is called its *footprint*. Within this footprint earth stations can communicate with each other via the satellite at any time during the day or night.

Large earth station dish antennae of up to 30 m diameter are used for telecommunications and TV distribution. Medium-sized earth stations use 3 to 5 m dishes for the transfer of commercial data such as telex and facsimile transmissions. Small dishes, up to about 80 cm diameter, are used for the reception of TV signals broadcast by satellite directly to individual homes.

Satellite television

Satellite television will undoubtedly bring about one of the biggest advances in home entertainment since broadcasting first began. In 1990 two English language satellite television stations began broadcasting to UK homes, Sky Television and British Satellite Broadcasting (BSB). These stations will, together, give access to some 20 different television channels in any suitably equipped home.

To receive either station, each house will require a dish aerial and a receiver. This receiver is a 'blackbox' about half the size of a video recorder which will turn the signals from the dish into sound and pictures on the television set.

Sky Television and BSB will broadcast from different satellites in difererent parts of the sky and will use different systems. Each will, therefore, require its own dish and receiver.

While initial satellite dish sales have been slow, commercial analysts anticipate that 10 million homes in the UK will have their own satellite TV system by the late 1990s. The installation of satellite dishes therefore represents a market with enormous growth potential for any electrical contractor or competent installer.

Satellite dish installation

The installation of a satellite dish would at first glance seem to require a degree from N.A.S.A. in mathematics, geometry and navigation! The installation instructions talk of azimuths, lines of longitude and latitude, orbits and inclinometers. The practical considerations are, in fact, much simpler. Installation is ultimately a matter of

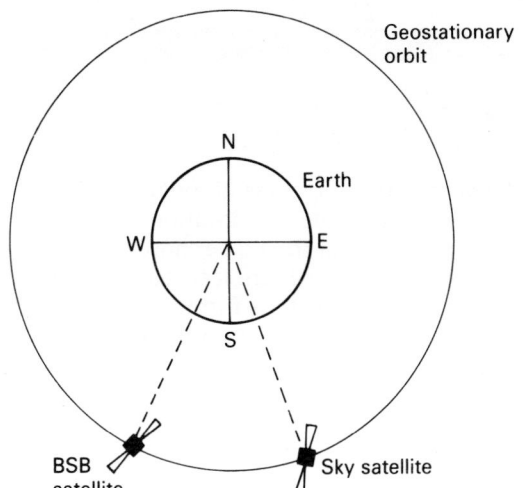

Figure 10.6 Approximate satellite positions

pointing the dish at the satellite and making sure it doesn't fall off the wall.

Site survey

The satellite is parked over the equator in a stationary orbit and, therefore, any dish antenna mounted in the UK must point south towards the satellite. The Sky satellite is to the east, or left of a reference line pointing true south and the BSB satellite is to the west or right of true south, as shown by Figure 10.6. The satellite dish must have a clear line of sight to the satellite position because microwaves will not pass through buildings, fences or trees. It must not be mounted above the line of the roof because this will probably break the local planning regulations.

Dish assembly and fixing

Upon arrival at the site, the dish should be assembled. The feedhorn assembly is usually held at the focal length of the dish by three rods. This distance is critical and must be assembled exactly as instructed by the manufacturer. When the dish has been assembled, check for dish warp. Dish warp and incorrect focal length are common causes of poor reception. The easiest way to check for dish warp is to look across the dish from the front edge to the back. Each edge should be perfectly aligned. However, if the dish is warped, first check the fixing of the support rods, and if these are correctly attached to a metal or aluminium dish, the distortion can be twisted out of the dish until the edges are aligned.

Fix the mounting brackets to the chosen position with Rawlbolts, coachscrews or number 12 wood screws into good-quality nylon wall plugs or solid timber as is appropriate. Wind gusts can greatly increase the forces acting upon the dish aerial and mounting bracket and, therefore, fixings must be secure. Fix the dish on to the mounting bracket and adjust the direction of the dish so that it points at the chosen satellite. The elevation angle can be measured with a protractor and plumb bob as shown in Figure 10.7.

Dish alignment

To point the dish at the chose satellite it must be tilted upwards (called the elevation angle) and

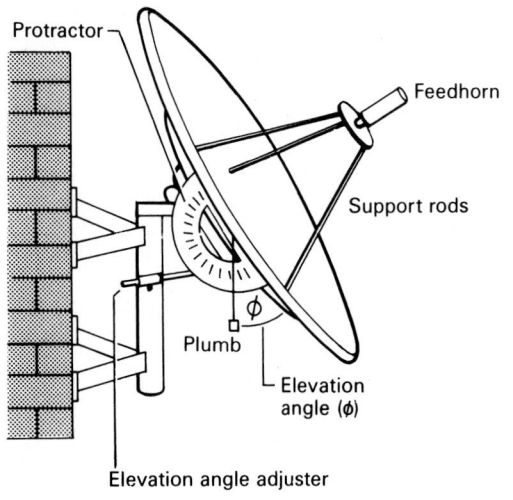

Figure 10.7 Setting up the elevation angle of a satellite dish using a protractor to measure the elevation angle ϕ

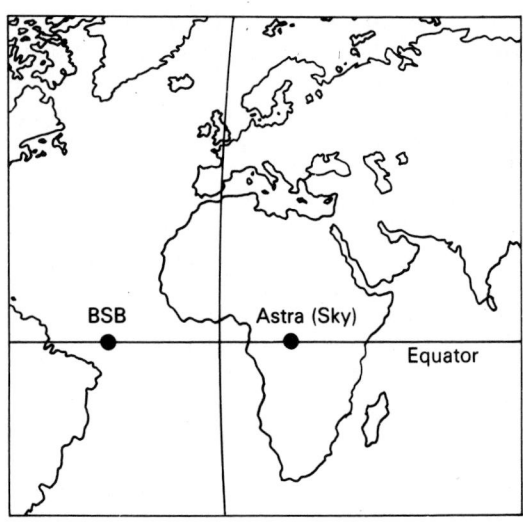

Figure 10.8 The position of two satellites in their geostationary orbits above the equator

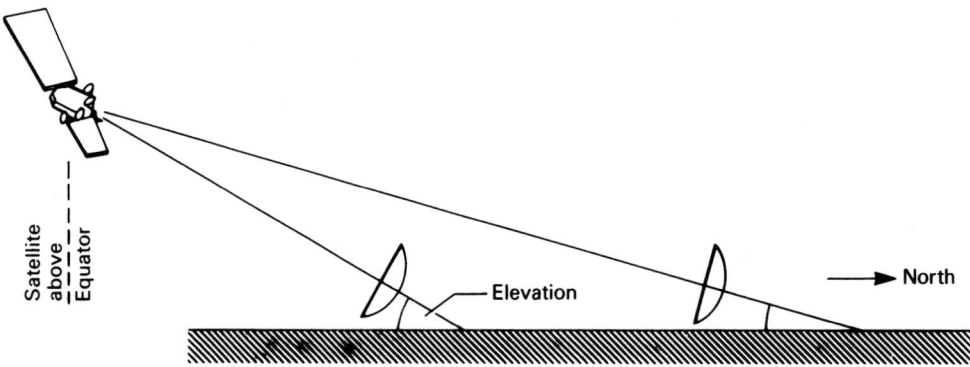

Figure 10.9 Dish elevation angle is less in the North than in the South

then rotated to the left of a line running true south for the Sky satellite or right of this line to pick up the BSB satellite. This is called the azimuth angle. Figure 10.8 shows the position of these two satellites in their stationary orbits above the equator.

Elevation angle

The dish elevation angle, that is the amount of tilt required for any site in the UK, can be found by considering Table 10.3. This table gives the elevation and azimuth angles for four different satellites and twenty-nine locations throughout the UK. When using this table, the location closest to the receiver site location should be chosen. The elevation angles in the north are less than those in the south because the north is further away from the equator and the position of the satellite as shown in Figure 10.9.

Azimuth angle

The dish azimuth angle is also given in Table 10.3 for various site locations.

Corrected azimuth angle

Having identified the azimuth angle for a specific satellite and site location the dish may be pointed in the correct direction with the aid of a compass. However, all magnetic compasses point to the magnetic north and magnetic south and the azimuth angles given in Table 10.3 are given for *true south*. True south is between 4° and 9° to the west of magnetic south at different parts of the UK as shown in Figure 10.10. The azimuth angles must, therefore, be corrected for the specific locations as follows:

Find the installation site location on the map shown in Figure 10.11 and

1. subtract the correction values from the azimuth angles given in Table 10.3 for the Sky satellite or
2. add the correction values to the azimuth angles given in Table 10.3 for the BSB satellite.

The azimuth angle will then be measured from the magnetic south indicated by the compass.

Example 1

A Sky dish antenna is to be installed in Manchester. The agreed place for the dish to be mounted is close to the eaves of a south-facing gable end. From this position there is an unrestricted view of the southern horizon. Determine the elevation and azimuth angles for the location.

Table 10.3 gives an azimuth angle of 26.05° east and an elevation angle of 25.79° for the Sky satellite. Figure 10.11 gives a correction angle of 6.5° for Manchester, which must be subtracted from the azimuth angle given in Table 10.3 for the Sky satellite.

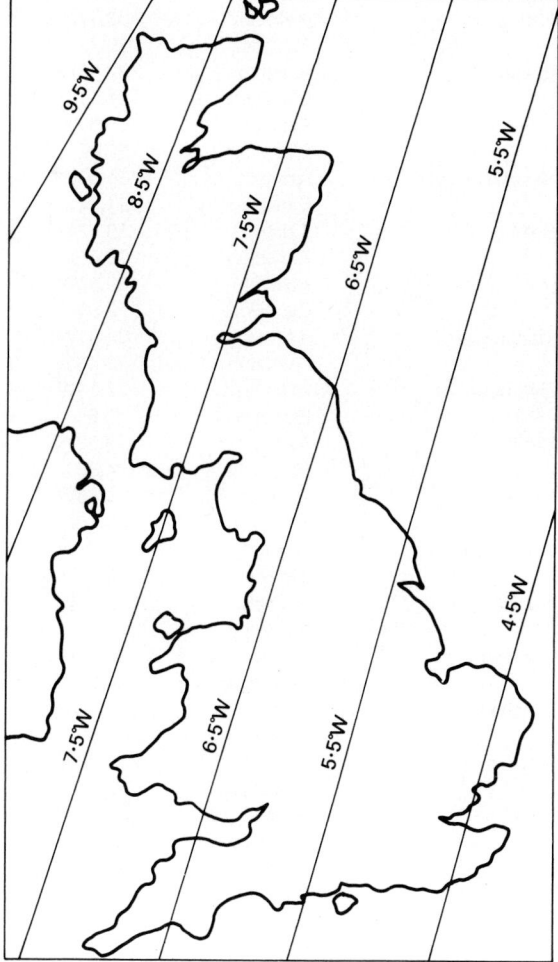

Figure 10.11 Amount of correction required to convert magnetic south to true South readings

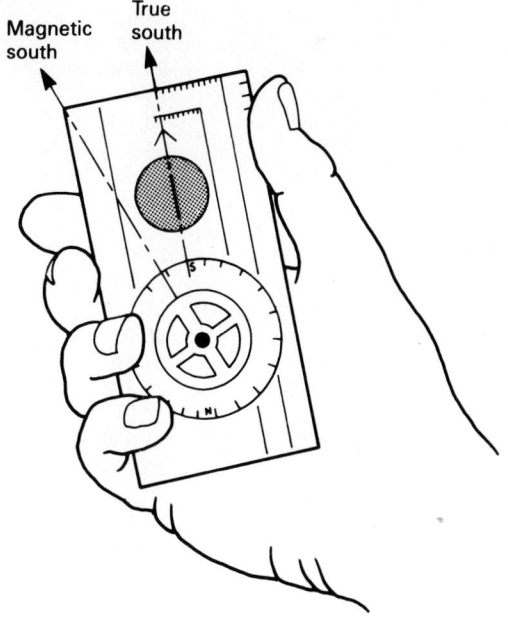

Figure 10.10 Compass readings showing true South and magnetic South

Table 10.3   Dish azimuth and elevation angles for various locations throughout the United Kingdom

Site locations		Satellite location (°)			
		ADD the correction angles given in Figure 10.11 to azimuth angles and swing dish to the RIGHT of magnetic south		SUBTRACT the correction angles given in Fig. 10.11 from azimuth angles and swing dish to the LEFT of magnetic south	
		BSB	Intelsat	Eutelsat	Astra (Sky)
John O'Groats	Azimuth	31.26W	27.42W	19.24E	25.61E
	Elevation	19.3	20.24	21.84	20.65
Wick	Azimuth	31.86W	28.03W	18.72E	27.5E
	Elevation	19.36	20.32	22.15	20.85
Inverness	Azimuth	30.88W	27.01W	20.22E	27.23E
	Elevation	20.57	21.54	22.91	21.49
Aberdeen	Azimuth	33.3W	29.47W	17.82E	24.91E
	Elevation	20.21	21.26	23.63	22.33
Edinburgh	Azimuth	32.46W	28.58W	19.33E	26.46E
	Elevation	21.61	22.66	24.62	23.18
Glasgow	Azimuth	31.34W	27.43W	20.56E	27.66E
	Elevation	22.02	23.04	24.5	22.98
Newcastle upon Tyne	Azimuth	34.53W	30.67W	17.65E	24.89E
	Elevation	21.94	23.09	25.93	24.54
Belfast	Azimuth	29.83W	28.85W	22.84E	29.94E
	Elevation	23.69	24.7	25.37	23.66
Hull	Azimuth	36.36W	32.51W	16.34E	23.71E
	Elevation	22.53	23.77	27.42	26.07
Manchester	Azimuth	34.31W	30.4W	18.73E	26.05E
	Elevation	23.46	24.65	27.3	25.79
Liverpool	Azimuth	33.51W	29.58W	19.65E	26.94E
	Elevation	23.8	24.97	27.23	25.66
Dublin	Azimuth	29.83W	25.8W	23.58E	30.73E
	Elevation	24.96	26.01	26.51	24.7
Galway	Azimuth	26.68W	22.59W	26.82E	33.85E
	Elevation	25.86	26.8	25.83	23.83
Great Yarmouth	Azimuth	38.97W	35.16W	14.05E	21.57E
	Elevation	22.65	24.01	28.9	27.66
Norwich	Azimuth	38.48W	34.65W	14.63E	22.14E
	Elevation	22.86	24.21	28.85	27.57
Birmingham	Azimuth	35.04W	31.12W	18.55 E	25.95E
	Elevation	24.21	25.45	28.39	26.85
Bedford	Azimuth	36.75W	32.86W	16.89E	24.37E
	Elevation	23.98	25.29	29.06	27.61
Ipswich	Azimuth	39.09W	35.27W	14.32E	21.89E
	Elevation	23.24	24.63	29.57	28.29
Cork	Azimuth	27.54W	23.41W	26.75E	33.85E
	Elevation	26.97	27.96	27.17	25.11
Swindon	Azimuth	35.54W	31.59W	18.62E	26.09E
	Elevation	24.99	26.28	29.39	27.81
London	Azimuth	37.39W	33.51W	16.55E	24.1E
	Elevation	24.34	25.69	29.75	28.3
Cardiff	Azimuth	34.01W	30.02W	20.35E	27.76E
	Elevation	25.6	26.84	29.17	27.47

Table 10.3   Dish azimuth and elevation angles for various locations throughout the United Kingdom (continued)

Site locations		Satellite location (°)			
		ADD the correction angles given in Figure 10.11 to azimuth angles and swing dish to the RIGHT of magnetic south		SUBTRACT the correction angles given in Fig. 10.11 from azimuth angles and swing dish to the LEFT of magnetic south	
		BSB	Intelsat	Eutelsat	Astra (Sky)
Bristol	Azimuth	34.69W	30.72W	19.63E	27.07E
	Elevation	25.4	26.66	29.33	27.68
Dover	Azimuth	39.14W	35.29W	14.91E	22.56E
	Elevation	24.2	25.63	30.56	29.2
Southampton	Azimuth	36.23W	32.29W	18.28E	25.82E
	Elevation	25.4	26.73	30.14	28.55
Brighton	Azimuth	37.61W	33.71W	16.77E	24.73E
	Elevation	24.96	26.34	30.47	28.98
Plymouth	Azimuth	33.31W	29.72W	21.83E	29.27E
	Elevation	26.93	28.17	30.02	28.17
Penzance	Azimuth	31.87W	27.77W	23.63E	31.02E
	Elevation	27.78	28.99	30.03	28.05
Landsend	Azimuth	31.6W	27.49W	23.88E	31.52E
	Elevation	27.82	29.01	29.92	27.93

The corrected azimuth angle $= 26.05° - 6.5° = 19.55°$. The Sky dish must, therefore, be elevated through an angle of 25.79° measured on a protractor as shown in Figure 10.7, and then turned left through 19.55° from the magnetic south indicated on a compass.

Example 2

A Sky dish aerial is to be installed on the south-facing wall of a London town house. The site survey has identified a suitable position. Calculate the elevation and azimuth angles for this location.

Table 10.3 gives an azimuth angle of 24.1° east and an elevation angle of 28.3° for the Sky satellite. Figure 10.11 gives a correction angle of 5° for London which must be subtracted from the azimuth angle given in Table 10.3 for the Sky satellite.

The corrected azimuth angle $= 24.1° - 5° = 19.1°$. The Sky dish must, therefore, be elevated through an angle of 28.3° and then turned left through 19.1° from the magnetic south indicated on a compass.

Example 3

A BSB antenna is to be installed on a suitable south-facing wall of a house in Manchester. Determine the elevation and azimuth angles for this location.

Table 10.3 gives an azimuth angle of 34.31° west and an elevation angle of 23.46° for the BSB satellite. Figure 10.11 gives a correction of 6.5° for Manchester which must be added to the azimuth angle given in Table 10.3 for the BSB satellite.

The corrected azimuth angle $= 34.31° + 6.5° = 40.81°$. The BSB dish must, therefore, be elevated through an angle of 23.46° and then turned *right* through an angle of 40.81° measured from the magnetic south indicated on a compass.

Fine tuning for maximum signal strength

Having adjusted the satellite dish for the correct elevation and azimuth angles, tighten the screws sufficiently to hold the dish still and run low-loss co-axial cable to the pre-tuned receiver unit. The signal transmissions from the satellite are polarised and the feedhorn assembly must be rotated for maximum signal strength.

Connect a signal strength meter (these cost about £50) or a TV monitor on a long lead and rotate the feedhorn and fine tune the elevation and azimuth angles for the best picture quality or maximum signal strength. It is important to tune for maximum signal strength. On a fine day the signal from the satellite is at its strongest but rain and snow will weaken the signal which may lead to problems later, such as sparklies. Finally, tighten up all nuts and bolts while continuing to monitor the picture quality or signal strength and maintaining the best results.

Practical considerations

Sparklies

Sparklies show as comet-shaped dots randomly distributed over the picture, white on dark areas and black on white areas. These are caused by a weak signal strength, and the elevation and azimuth angles should be checked if this problem is evident.

Spiders

Spiders climb inside the feedhorn, particularly in the autumn. This weakens the signal strength.

Co-axial cable

If adjustment or cleaning does not remove sparklies, then the dish itself may be too small or the co-axial down lead cable may be mismatched or too long. If the output impedance of the dish is 75 Ω, then 75 Ω impedance co-axial cable must be used. Avoid kinks or sharp bends in the co-axial down lead.

Termination

Termination of the cable at the feedhorn is best done by F-type crimp connectors. The crimp tool compresses the connector on to the jacket equally on all sides and helps to waterproof the joint.

Water ingress

Water ingress is a problem where the cable terminates at the feedhorn. Rubber boots, shrouds and heat-shrink sleeves are available and will help. The cable connector and cable should be wrapped with rubber tape at the feedhorn. PVC tape does not offer sufficient weatherproofing for this particular job because it hardens in cold weather.

Regulations

Local planning laws will allow only one dish on each building and this must be less than 90 cm in diameter. The dish must not be mounted above the roof line of the house. Installations which require bigger dishes must first obtain permission from the Local Planning Authority. An ordinary television licence will be required by satellite dish owners in addition to any monthly subscriptions to the satellite broadcasting company.

Cable television

An alternative to installing a separate roof-top receiving antenna on each house, is to supply many houses with television pictures by underground cables fed from one remote ground station.

Cable companies receive satellite programmes in addition to BBC and ITV programmes and send them straight into the home via a co-axial or fibre-optic cable which plugs straight into the existing TV set.

Since private companies will provide the cable TV system, the customer will pay the operator a monthly fee to view their chosen channels. At the moment it is only available in certain areas and for economic reasons will only become available to viewers in areas of high population density.

Computer supplies

Many computer, data processing and communication systems are sensitive to variations or distortions in the a.c. mains supply. Distortions, interference, pulses or spikes on the mains are collectively called 'noise', which can be caused by switching inductive circuits on or off, motor control equipment, brushgear, sparking of commutator motors, welders or thyristor switching of speed controllers. Variations outside very narrow limits can cause computers to 'crash' or provoke computer errors. For these reasons it is sometimes necessary to supply computer systems with a 'clean' supply or 'suppress' the possible sources of noise.

A clean supply is a separate supply which feeds items of computer equipment. This is usually fed from a point as close as possible to the LV supply source. A clean earth is also taken from the supply point, usually as one core of the feeder cable. Final distribution of the 'clean' supply is then by standard wiring circuits. Alternatively the supply to the computer equipment can be cleaned by suitable input filters. Mains suppression filters are available in many forms from any reputable supplier.

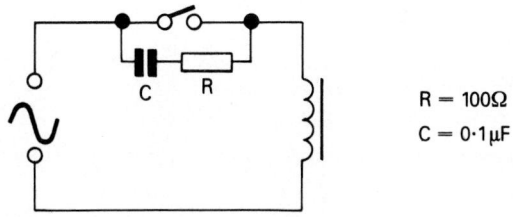

Figure 10.12 A simple noise suppressor

$$R = 100\Omega$$
$$C = 0.1\mu F$$

Noise generated by switch contact arcing can be suppressed by either connecting an R-C circuit across the contacts as shown in Figure 10.12 or by connecting a Metal Oxide Varistor (MOV) across the contacts. An MOV looks like a capacitor but behaves like a resistor, dissipating the energy contained in the noise. This has the additional benefit of increasing switch life by reducing arc damage at the contacts.

Computer networks

The electronic typewriter of the '80s is being joined by the desktop personal computer (PC) in the office of the '90s. If a number of personal computers are to 'talk' to each other or share access to centrally held information, records or data, then they must be connected together in a 'network'.

Local area networks of computers (LANs)

When a number of computers are connected together in the same room, office or building, they are called *local area networks* (LANs). LANs are often used in the office environment where a number of personal computers share software and central file storage facilities. A typical example of a LAN installation might be a word-processing pool where files of standard letters, customer and product details are held centrally and accessed by a number of work stations using the same word-processing software. Insurance offices and travel agents use LANs so that a number of clerical staff can access information held centrally relating to policies or holidays, issue cover notes and make bookings. LANs are also used effectively in retail outlets to maintain a record of all sales. The cash tills are microcomputers which update a central file every time an item is sold which can, in turn, update the stock control records.

Types of network

LANs may be wired in different configurations to suit a particular application. A *star* network is used when a number of users require access to a central store of information. The data base forms the centre point of the star, and the work stations radiate outwards.

When workstations need to communicate, for example when being used for electronic mail between offices, a *mesh* connected LAN might be used as shown in Figure 10.13. A *bus* network is

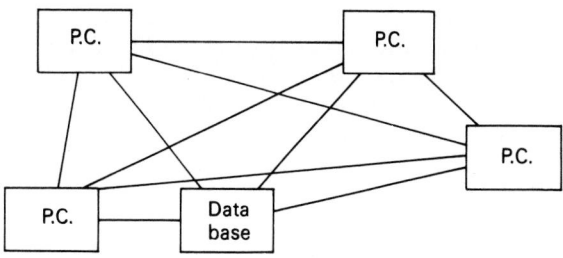

Figure 10.13 A fully interconnected network

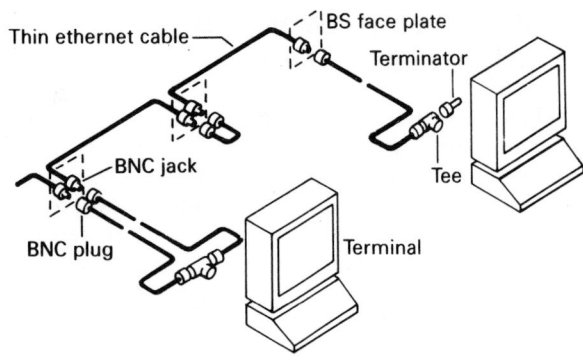

Figure 10.14 LAN cable layout

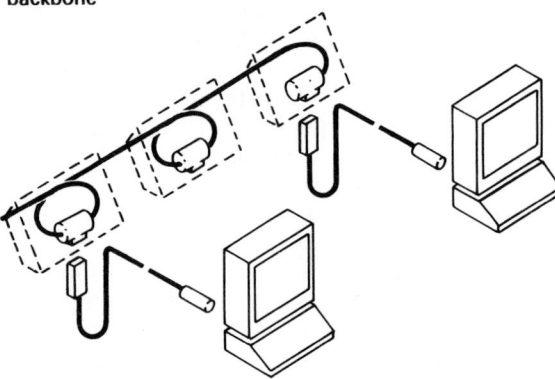

This easily installed tap for Ethernet/
IEEE 802.3 10 Base 2 network provides
pluggable access to the network
backbone

SIMPLE INSTALLATION
Installs quickly on IEE 802.3 10 Base 2
50-ohm cable without special tools

INTERNAL SWITCHING
Maintains series circuit so that drop
cable connection and dis-connection
does not disrupt operating network

Figure 10.15 LAN installation using the AMP Thinnet
cable

one which is widely used in offices where each workstation is connected to a central communications highway. Many network systems are commercially available. *Ethernet* is a well-known standard bus network developed by *Xerox*. This involves looping a 50 Ω co-axial cable around the LAN and terminating the workstations with BNC plug connectors. When workstations are removed the BNC plugs must be linked to maintain the continuity of the LAN as shown in Figure 10.14.

The *Amp Thinnet* bus network is a development of the *Ethernet* network. This uses thinner RG58 (50 Ω) co-axial cable and loop in socket termination (called TAPS) which contains an internal switch so that the LAN is not broken when a workstation is disconnected. This system can feed up to ten active workstations with a total bus length of 185 metres. The layout is shown in Figure 10.15. The network cables are sometimes referred to as the *transmission medium* which might be co-axial cable, twisted-pair cable or flat ribbon cable. The cables can be terminated at each workstation in BNC sockets, 'D' connectors or *Taps* as shown in Figure 10.16.

Whatever the transmission medium, many cables will require containing in a suitable enclosure. Companies now invest heavily in office equipment and most are interior design conscious, recognising that the office is an important part of their public image. Making the network installation look good is, therefore, just as important as providing an efficient and flexible system. The power cables and network cables will also require segregating, not only because of IEE Regulation 525-1, but to prevent mains-carried 'noise' on the

computer network. These requirements are usually met by a multicompartment trunking system installed overhead or underfloor or run around the perimeter of the room. A skirting trunking of the type shown in Figure 10.17 can provide an attractive solution and be fitted at floor or desktop level.

Security systems

The installation of security alarm systems in this country is already a multi-million-pound business and yet it is also a relatively new industry. As society becomes increasingly aware of crime prevention, it is evident that the market for security systems will expand.

Not all homes are equally at risk, but all homes have something of value to a thief. Properties in cities are at highest risk, followed by homes in towns and villages and at least risk are homes in rural areas. A nearby motorway junction can, however, greatly increase the risk factor. Flats and maisonettes are the most vulnerable, with other

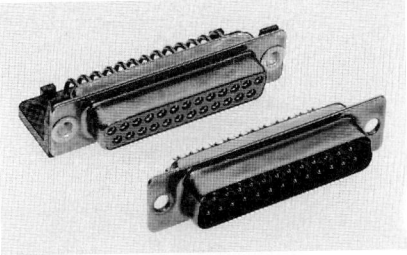

D-Connector

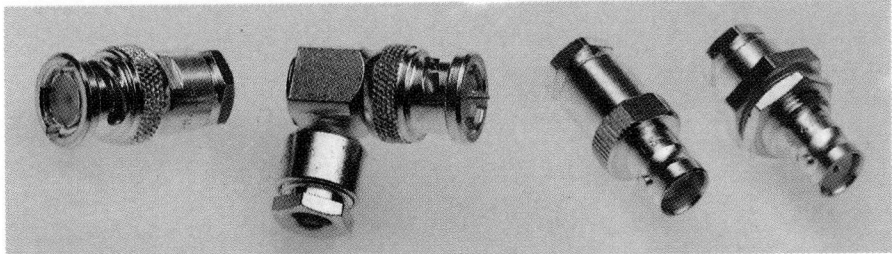

BNC Bayonet locking connectors

Figure 10.16 LAN network cable terminations

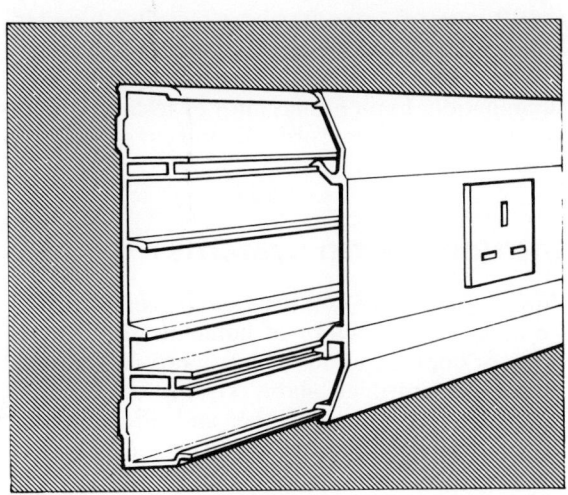

Figure 10.17 Trunking system suitable for LAN distribution

types of property at roughly equal risk. Most intruders are young, fit and foolhardy opportunists. They ideally want to get in and away quickly but, if they can work unseen, they may take a lot of trouble to gain access to a property by, for example, removing the putty from a window.

Most intruders are looking for portable and easily saleable items such as video recorders, television sets, home computers, jewellery, cameras, silverware, money, cheque books or credit cards. The Home Office have stated that only 7% of homes are sufficiently protected against intruders although 75% of householders believe they are secure. Taking the most simple precautions will reduce the risk, installing a security system can greatly reduce the risk of a successful burglary.

Security lighting

Security lighting is the first line of defence in the fight against crime. 'Bad men all hate the light and avoid it, for fear that their practices might be shown up' (John 3:20). A recent study carried out by Middlesex Polytechnic has shown that in two London boroughs the crime figures were reduced by improving the lighting levels. Police forces

Figure 10.18 Security lighting reduces crime

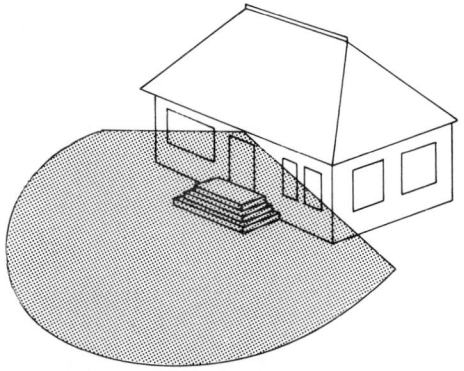

Figure 10.19 PIR detector, field of detection

agree that homes which are externally well illumin-ated are a much less attractive target for the thief.

Security lighting installed on the outside of the home is activated by external detectors. These detectors sense the presence of a person outside the protected property and additional lighting is switched on. This will deter most potential intrud-ers whilst also acting as courtesy lighting for visitors.

PIR detectors

Passive infra-red detector units allow a householder to switch on lighting units automatically whenever the area covered is approached by a moving body whose thermal radiation differs from the back-ground. This type of detector is ideal for driveways or dark areas around the protected property. It also saves energy because the lamps are only switched on when someone approaches the pro-tected area. The major contribution to security lighting comes from the 'unexpected' high-level illumination of an area when an intruder least

expects it. This surprise factor often encourages the potential intruder to 'try next door'.

Passive infra-red detectors are designed to sense heat changes in the field of view dictated by the lens system. The field of view can be as wide as 180° as shown by the diagram in Figure 10.19. Many of the 'better' detectors use a split lens system so that a number of beams have to be broken before the detector switches on the security lighting. This capability overcomes the problem of false alarms and a typical PIR is shown in Figure 10.20.

PIR detectors are often used to switch tungsten halogen floodlights because, of all available lumi-naires, tungsten halogen offers instant high-level illumination. Light fittings must be installed out of reach of an intruder in order to prevent sabotage of the security lighting system.

Intruder alarm systems

Alarm systems are now increasingly considered to be an essential feature of home security for all types of homes and not just property in high-risk areas. An intruder alarm system serves as a deterrent to a potential thief and often reduces home insurance premiums. In the event of a burglary they alert the occupants, neighbours and officials to a possible criminal act and generate fear and uncertainty in the mind of the intruder which encourages a more rapid departure. Intruder alarm systems can be broadly divided into three categories. Those which give perimeter protec-tion, space protection, or trap protection. A

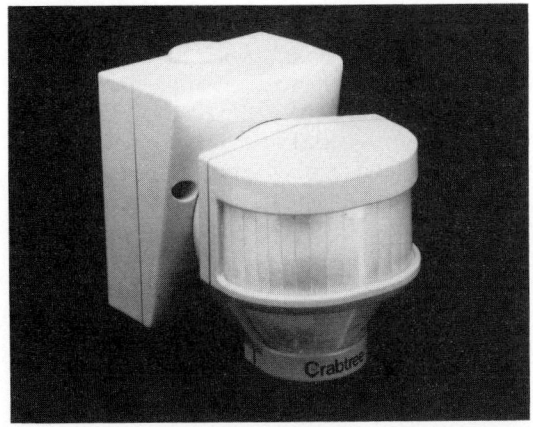

Figure 10.20 The Crabtree Minder – a typical PIR detector

Figure 10.21 Proximity switches for perimeter protection

system can comprise one or a mixture of all three categories.

Perimeter protection

A perimeter protection system places alarm sensors on all external doors and windows so that an intruder can be detected as he attempts to gain access to the protected property. This involves fitting proximity switches to all external doors and windows.

Space protection

A movement or heat detector placed in a room will detect the presence of anyone entering or leaving that room. Passive infra-red detectors and ultrasonic detectors give space protection. Space protection does have the disadvantage of being triggered by domestic pets but it is simpler and, therefore,

cheaper to install. Perimeter protection involves a much more extensive and, therefore, expensive installation, but is easier to live with.

Trap protection

Trap protection places alarm sensors on internal doors and pressure pad switches under carpets on through routes between, for example, the main living area and the master bedroom. If an intruder gains access to one room he cannot move from it without triggering the alarm.

Proximity switches

These are designed for the discreet protection of doors and windows. They are made from moulded plastic and are about the size of a chewing-gum packet, as shown in Figure 10.21. One moulding contains a reed switch, the other a magnet, and when they are placed close together the magnet maintains the contacts of the reed switch in either an open or closed position. Opening the door or window separates the two mouldings and the switch is activated, triggering the alarm.

Passive infra-red detector

These are activated by a moving body which is warmer than the surroundings. The PIR shown in Figure 10.22 has a range of 12 m and a detection zone of 110° when mounted between 1.8 m and 2 m high.

Figure 10.22 PIR intruder alarm detector

Ultrasonic detector

The ultrasonic motion detector is able to recognise the difference between random motion and intruder movement in a room. They are usually mounted in the corner of a room and have a detection range of 9 m. Figure 10.23 shows a typical ultrasonic detector.

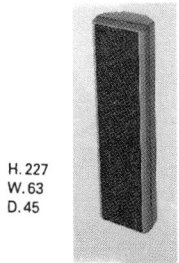

H. 227
W. 63
D. 45

Figure 10.23 Ultrasonic motion detector

Pressure pads

Pressure pad switches, such as those shown in Figure 10.24, are placed under the carpet close to a door. Treading on the carpet activates the switch and the alarm system.

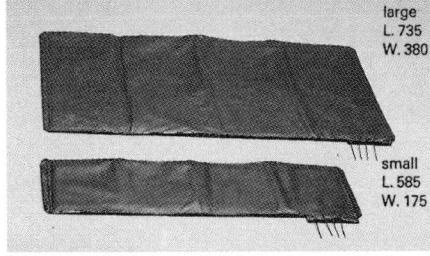

large
L. 735
W. 380

small
L. 585
W. 175

Figure 10.24 Pressure pad intruder alarm detectors

Intruder alarm sounders

Alarm sounders give an audible warning of a possible criminal act. Bells or sirens enclosed in a waterproof enclosure, such as shown in Figure 10.25, are suitable. It is usual to connect two sounders on an intruder alarm installation, one inside to make the intruder apprehensive and anxious, hopefully encouraging a rapid departure from the premises, and one outside. The outside sounder should be displayed prominently since the installation of an alarm system is thought to deter the casual intruder and a ringing alarm encourages neighbours and officials to investigate a possible criminal act.

Figure 10.25 Intruder alarm sounder

Control panel

The control panel such as that shown in Figure 10.26 is at the centre of the intruder alarm system. All external sensors and warning devices radiate

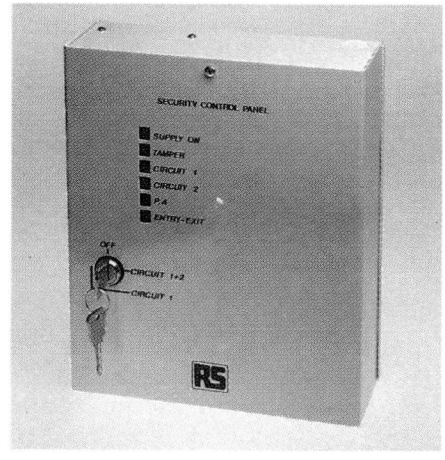

Figure 10.26 Intruder alarm control panel

from the control panel. The system is switched on or off at the control panel using a switch or coded buttons. To avoid triggering the alarm as you enter or leave the premises, there are exit and entry delay times to allow movement between the control panel and the door.

Supply

The supply to the intruder alarm system must be secure and this is usually achieved by an a.c. mains supply and battery back-up. Nickel cadmium rechargeable cells are usually mounted in the sounder housing box.

Design considerations

It is estimated that there is now a one-in-twenty chance of being burgled but the installation of a security system does deter a potential intruder. Every home in this country will almost certainly contain electrical goods, money or valuables of value to an intruder. Installing an intruder alarm system tells the potential intruder that you intend to make his job difficult, which in most cases encourages him to look for easier pickings.

The type and extent of the intruder alarm installation and, therefore, the cost will depend upon many factors including the type and position of the building, the contents of the building, the insurance risk involved and the peace of mind offered by an alarm system to the owner or occupier of the building.

The designer must ensure that an intruder cannot sabotage the alarm system by cutting the wires or pulling the alarm box from the wall. Most systems will trigger if the wires are cut and sounders should be mounted in any easy-to-see but difficult-to-reach position.

Intruder alarm circuits are category 2 circuits and should, therefore, be segregated from mains supply cables or insulated to the highest voltage present if run in a common enclosure with category 1 cables (IEE Regulation 525-1).

APPENDICES

Appendix A: Obtaining information and components

For local suppliers, you should consult your local telephone directory. However, the following companies distribute electrical and electronic components throughout the UK. In most cases, telephone orders received before 5 pm can be dispatched the same day.

Electromail (R. S. mail order business), P.O. Box 33, Corby, Northants NN17 9EL Tel. (0536) 204555.

Farnell Electronic Components, Canal Road, Leeds LS12 2TU Tel. (0532) 636311.

Maplin Electronics, P.O. Box 777, Rayleigh, Essex SS6 8LV Tel. (0702) 552961.

Rapid Electronics Ltd, Heckworth Close, Severalls Industrial Estate, Colchester, Essex CO4 4TB Tel. (0206) 751166.

R.S. Components Ltd, P.O. Box 99, Corby, Northants NN17 9RS Tel. (0536) 201234.

Verospeed Electronic Components, Boyatt Wood, Eastleigh, Hants SO5 4ZY Tel. (0703) 644555.

Appendix B: Abbreviations, symbols and codes

Abbreviations used in electronics for multiples and sub-multiples

T	tera	10^{12}
G	giga	10^{9}
M	mega or meg	10^{6}
k	kilo	10^{3}
d	deci	10^{-1}
c	centi	10^{-2}
m	milli	10^{-3}
μ	micro	10^{-6}
n	nano	10^{-9}
p	pico	10^{-12}

Terms and symbols used in electronics

Term	Symbol
Approximately equal to	$\simeq$
Proportional to	$\propto$
Infinity	∞
Sum of	Σ
Greater than	$>$
Less than	$<$
Much greater than	$\gg$
Much less than	$\ll$
Base of natural logarithms	e
Common logarithms of x	$\log x$
Temerature	θ
Time constant	T
Efficiency	η
Per unit	p.u.

Electrical quantities and units

Quantity	Quantity symbol	Unit	Unit symbol
Angular velocity	ω	radian per second	rad/s
Capacitance	C	farad	F
		microfarad	μF
		picofarad	pF
Charge or quantity of electricity	Q	coulomb	C
Current	I	ampere	A
		milliampere	mA
		microampere	μA
Electromotive force	E	volt	V
Frequency	f	hertz	Hz
		kilohertz	kHz
		megahertz	MHz
Impedance	Z	ohm	Ω
Inductance, self	L	henry (plural, henrys)	H
Inductance, mutual	M	henry (plural, henrys)	H
Magnetic field strength	H	ampere per metre	A/m
Magnetic flux	θ	weber	Wb
Magnetic flux density	B	tesla	T
Potential difference	V	volt	V
		millivolt	mV
		kilovolt	kV
Power	P	watt	W
		kilowatt	kW
		megawatt	MW
Reactance	X	ohm	Ω
Resistance	R	ohm	Ω
		microhm	$\mu\Omega$
		megohm	MΩ
Resistivity	ρ	ohm metre	Ωm
Wavelength	λ	metre	m
		micrometre	μm

Capacitor values – conversion table

Capacitance (Picofarad PF)	Capacitance (Nanofarad NF)	Capacitance (Microfarad μF)	Capacitance code
10	0.01		100
15	0.015		150
47	0.047		470
82	0.082		820
100	0.1		101
330	0.33		331
470	0.47	0.00047	471
1000	1.0	0.001	102
1500	1.5	0.0015	152
2200	2.2	0.0022	222
4700	4.7	0.0047	472
6800	6.8	0.0068	682
10000	10	0.01	103
22000	22	0.022	223
47000	47	0.047	473
100000	100	0.1	104
220000	220	0.22	224
470000	470	0.47	474

capacitance code
First two digits significant figures; third is number of zeros. Value
given in pF

Suffixes used with semiconductor devices

Many semiconductor devices are available with suffix
letters after the part number, i.e. BC108B, C106D, TIP31C.
The suffix isused to indicate a specific parameter
relevant to the device – some examples are shown below.

Thyristors, triacs, power rectifiers
Suffix indicates voltage rating, e.g. TIC106D indicates
device has a 400 volt rating. Letters used are:

Q = 15 Volts	B = 200 Volts	M = 600 Volts
Y = 30 Volts	C = 300 Volts	S = 700 Volts
F = 50 Volts	D = 400 Volts	N = 800 Volts
A = 100 Volts	E = 500 Volts	

Small signal transistors
Suffix indicates h_{FE} range, e.g. BC108C
A = h_{FE} of 125–260
B = h_{FE} of 240–500
C = h_{FE} of 450–900

Power transistors
Suffix indicates voltage, e.g. TIP32C
No suffix = 40 volts
A = 60 volts
B = 80 volts
C = 100 volts
D = 120 volts

Resistor and capacitor letter and digit code (BS 1852)

Resistor values are indicated as follows:

0.47 Ω	marked	R47	100 Ω	marked	100R
1 Ω		1R0	1 kΩ		1K0
4.7 Ω		4R7	10 kΩ		10K
47 Ω		47R	10 MΩ		10M

A letter following the value shows the tolerance.
F = ± 1%; G = ±2%; J = ±5%; K = ±10%; M = ±20%;
R33M = 0.33 Ω ± 20%; 6K8F = 6.8 kΩ ± 1%.

Capacitor values are indicated as:

0.68 pF	marked	p68	6.8 nf	marked	6n8
6.8 pf		6p8	1000 nF		1μ0
1000 pF		1n0	6.8 μF		6μ8

Tolerance is indicated by letters as for resistors. Values up to 999 pF are marked in pF, from 1000 pf to 999 000 pF (= 999 nF) as nF (1000 pF = 1 nF) and from 1000 nF (= 1 μF) upwards as μF.
Some capacitors are marked with a code denoting the value in pF (first two figures) followed by a multiplier as a power of ten ($3 = 10^3$). Letters denote tolerance as for resistors but C = ±0.25 pf.
E.g. 123J = 12 pF × 10^3 ± 5% = 12000 pF (or 0.12 μF).

Appendix C: Greek symbols

Greek letters used as symbols in electronics

Greek letter	Capital (used for)	Small	(used for)
Alpha	–	α	(angle, temperature coefficient of resistance, current amplification factor for common-base transistor)
Beta	–	β	(current amplification factor for common-emitter transistor)
Delta	Δ (increment, mesh connection	δ	(small increment)
Epsilon	–	ε	(permittivity)
Eta	–	η	(efficiency)
Theta	–	θ	(angle, temperature)
Lambda	–	λ	(wavelength)
Mu	–	μ	(micro, permeability, amplification factor)
Pi	–	π	(circumference/diameter)
Rho	–	ρ	(resistivity)
Sigma	Σ (sum of)	σ	(conductivity)
Phi	θ (magnetic flux)	ϕ	(angle, phase difference)
Psi	Ψ (electric flux)	–	
Omega	Ω (ohm)	ω	(solid angle, angular velocity, angular frequency)

Appendix D: Battery information

1. Types of battery

Alkaline primary cells

These cells and batteries offer very long service life compared with Leclanché types in equipments having high current drains. In addition these cells have very low self-discharge currents and are completely sealed.
Available in sizes AAA, AA, C, D and PP3.

Silver/mercuric oxide primary cells

These button cells are suitable for use in calculators, small tools, cameras, clocks, watches etc. They may often be used as replacements for previously fitted alkaline manganese button cell types. Supplied in boxes of individual blister packs.
Available in six of the most popular sizes.

Ni-Cad sintered cells

Applications where extreme ruggedness and/or high peak currents are required. In addition these cells offer very long service life and can be electrically misused without damage.
Available in sizes N, AAA, AA, C, D and PP9.

Ni-Cad high-temperature sintered cells

Primarily for use in emergency lighting installations these cells and batteries are particularly suitable for charging and discharging at elevated temperatures. Other applications include alarm control panels and emergency and standby areas where higher ambient temperatures are experienced.
Available as single D cells and 3 × D cell battery packs.

Note: sintered cells

Have fairly high self-discharge currents and are therefore not suitable for equipments which have to be operational without recharging, after being left unattended for long periods of time.

Ni-Cad mass plate cells

Applications where small size and ruggedness are required. These cells have low self-discharge currents and are ideal in many small portable equipments.
A range of sizes including PP3.

2. Battery rating and storage

Battery type and Stock Nos.	Ratings			Storage		
	Voltage		Capacity	Shelf life (see note 1)	Storage life	Storage temp.
	Fully charged	Discharged				
Alkaline Cells						
591-657 AAA	1.50 V	0.90 V	0.7 Ah			
591-225 AA	1.50 V	0.90 V	1 Ah	24	24	
591-231 C	1.50 V	0.90 V	4 Ah	months	months	−20°C + 50°C
591-247 D	1.50 V	0.90 V	8 Ah	@T_a = 20°C	@T_a = 20°C	
591-792 PP3	9.00 V	5.40 V	0.4 Ah		(see note 2)	
Silver Oxide Cells						
592-082 SR41	1.55 V	1.2 V	38 mAh	24 months	24 months	
592-098 SR43	1.55 V	1.2 V	120 mAh	@T_a = 20°C	@T_a = 20°C	−10°C + 25°C
592-105 SR44	1.55 V	1.2 V	140 mAh	to 90%	to 90%	recommended
592-111 SR54	1.55 V	1.2 V	80 mAh	capacity	capacity	
Mercuric Oxide Cells						
592-127 PX/RM 625	1.35 V	0.9 V	350 mAh	24 months	24 months	
592-133 PX/RM 400	1.35 V	0.9 V	70 mAh	@T_a = 20°C	@T_a = 20°C	−10°C + 25°C
				to 90%	to 90%	recommended
				capacity	capacity	
Ni-Cad Sintered Cells			(5hr discharge rate)			
592-026 N	1.24–1.27 V	1.00 V	150 mAh			
591-146 AAA	1.24–1.27 V	1.00 V	180 mAh	120 days		
591-051 AA	1.24–1.27 V	1.00 V	500 mAh	T_a = 0°C		
591-045 C	1.24–1.27 V	1.00 V	2 Ah	40 days	>5 years	−40°C + 60°C
591-039 D	1.24–1.27 V	1.00 V	4 Ah	T_a = 20°C	(see note 2)	
591-095 PP9	8.68–8.90 V	7.00 V	1.2 Ah	11 days		
				T_a = 40°C		
Ni-Cad High Temp Cells			(5hr discharge rate)			
592-032 D	1.24–1.27 V	1.00 V	4 A			−45°C + 65°C
592-048 3 × D, Stick	3.72–3.81 V	3.00 V	4 Ah	55 days	>5 years	possible
592-054 3 × D, Plate	3.72–3.81 V	3.00 V	4 Ah	T_a = 20°C	(see note 2)	0°C + 45°C
						recommended
Ni-Cad Mass Plate			(10 hr discharge rate)			
591-477 PCB Battery	3.72–3.81 V	3.00 V		26 months		Max limits
591-089 PP3	8.70–8.90 V	7.00 V	100 mAh	T_a = 0°C	>5 years	−40°C + 50°C
591-168 Button Cell	1.24–1.27 V	1.00 V	110 mAh	10 months	(see note 2)	0°C + 45°C
591-174 Button Cell	1.24–1.27 V	1.00 V	170 mAh	T_a = 20°C		recommended
591-180 Stack	8.70–8.90 V	7.00 V	280 mAh	1 month		
591-196 Stack	6.00–6.20 V	5.00 V	170 mAh	T_a = 40°C		
			280 mAh			

Note:
1. Period after which only 60% of the stated capacity is obtainable.

3.   Charging information for rechargeable batteries

Battery type and stock numbers		Charge mode	Charge rate Continuous	Max	Charge ∅ (note 1)	Temp range	Stock numbers for suitable RS chargers
Alkaline Cells							
591-657	AAA						
591-225	AA			Not			
591-231	C			rechargeable			
591-247	D						
591-792	PP3						
Silver Oxide Cells							
592-082	SR41			Not			
592-098	SR43			rechargeable			
592-105	SR44						
592-111	SR54						
Mercuric Oxide Cells							
592-127	PX/RM 625						
592-133	PX/RM 400			Not rechargeable			
Ni-Cad Sintered Cells							
592-026	N		15 mA	2C			
591-146	AAA		20 mA	2C	60% @ C/8		A wide range of suitable RS
591-051	AA	Series	66 mA				chargers is available.
591-045	C	constant	250 mA	10C	80% @ C/2	+ 10°C + 45°C	Please refer to current RS
591-039	D	current	500 mA				catalogue.
591-095	PP9		100 mA	2C	90% @ C		(note 3)
Ni-Cad High Temp. Cells							
592-032	D	Series	500 mA			(see note 2)	591-067, 591-714
592-048	3 × D, Stick	constant	500 mA	2C	84%	−20°C + 65°C	591-067
592-054	3 × D, Plate	current	500 mA			(reduced spec)	591-067
						+ 10°C + 35°C	
						(full spec)	
Ni-Cad Mass Plate							
591-477	PCB Battery		1.0 mA				–
591-089	PP3	Series	1.1 mA			Max limits	591–152
591-168	Button Cell	constant	1.7 mA	C/10	72%	0°C + 45°C	–
591-174	Button Cell	current	2.8 mA			+ 10°C + 35°C	–
591-180	Stack		1.7 mA			recommended	–
591-196	Stack		2.8 mA				–

Notes:

1. $\varnothing = \dfrac{\text{energy stored in the battery}}{\text{energy supplied to the battery}}$

2. At temperatures below 0°C charge current is limited to 120 mA and voltage to a maximum of 1.55 V/cell.

3. R.S. Components Ltd. (See Appendix A for further information)

4. Discharge information

Battery type and stock numbers		Discharge temp	I max (see note 1)	Cyclic life	Standby life	R_{int} (dc)	R_{int} (ac) 50 Hz
Alkaline Cells							(see note 3)
591-657	AAA						175 mΩ
591-225	AA						160 mΩ
591-231	C	−20°C + 50°C	0.9 A	1 cycle	2 years		130 mΩ
591-247	D						120 mΩ
591-792	PP3						2.7 Ω
Silver Oxide Cells			(see note 9)				
592-082	SR41		5/10 mA			8 Ω	
592-098	SR43		10/50 mA			6 Ω	
592-105	SR44	−10°C + 60°C	10/50 mA	1 cycle	2 years	4 Ω	
592-111	SR54		10/50 mA			10 Ω	
Mercuric Oxide Cells			(see note 9)				
592-127	PX/		5/10 mA			2 Ω	
	RM625	−10°C + 60°C	0.1/0.5 mA	1 cycle	2 years	5 Ω	
592-133	PX/						
	RM400						
Ni-Cad Sintered Cells							
592-026	N		0.9 A			105 mΩ	
591-146	AAA		1.0 A			80 mΩ	
591-051	AA		35 A	700–1000		20 mΩ	
591-045	C	−30°C + 50°C	70 A	cycles	7 years	8 mΩ	
591-039	D		110 A			5 mΩ	
591-095	PP9		2.4 A				
Ni-Cad High Temp Cells							(see note 3)
		(see note 7)					
592-032	D	−40°C + 65°C				6.5 mΩ	3.75 mΩ
592-048	3 × D.	(reduced spec)	24 A	800–1000	4–6 years	19.5 mΩ	11.25 mΩ
	Stick	−20°C + 45°C		cycles		19.5 mΩ	11.25 mΩ
592-054	3 × D.	(full spec)					
	Stick						
Ni-Cad Mass Plate		(see note 8)					(see note 3)
591-477	PCB	Max limits	180 mA			1.5 Ω	930 mΩ
	Battery	−20°C + 50°C	180 mA			3.5 Ω	2.17 Ω
591-089	PP3		300 mA	300	5 years	375 mΩ	190 mΩ
591-168	Button	0°C + 45°C	410 mA	cycles		200 mΩ	100 mΩ
	Cell	recommended	200 mA			1.87 Ω	850 mΩ
591-174	Button		410 mA			1.4 Ω	700 mΩ
	Cell						
591-180	Stack						
591-196	Stack						

Notes:

3. At 1 kHz.
7. A maximum of 75°C is permissible for up to 24 hours.
8. At temperatures below 0°C maximum discharge is C/2.
9. Continuous/Pulsed recommended maximum discharge currents.

Appendix E: Small signal diodes

Signal diodes

Package

Axial lead
a) L = 4.25, Dia = 1.85
b) L = 5.2, Dia = 2.7
d) L = 3.81, Dia = 1.71
e) L = 7.6, Dia = 2.7
f) L = 2.6, Dia = 1.7

Coloured band indicates cathode

TO-18

(k) TO-92

g) k — a
h) a case k
j) a — k

Application Code:
○ General Purpose — ◇ Switching — ● High speed
▽ VHF Tuner — □ Low leakage/low capacitance

Package	VRRM max (V)	IF(AV) max (mA)	VF max (V)	IF (mA) @	App'n Code	Order Code
(e)	100	140	0.8	250	○	AAZ15 ■
(e)	75	140	0.8	250	○	AAZ17 ■
(a)	30	100	1.1	100	○	BA317
(a)	35	100	1.2	100	◇	BA482
(f)	125	225	1.0	200	─	BAS45
(a)	100	100	0.45	1	○	BAT41 ◆
(a)	30	100	0.4	10	○	BAT42 ◆
(a)	30	100	0.45	15	○	BAT43 ◆
(a)	100	150	0.45	10	○	BAT46 ◆
(a)	20	350	0.4	10	◇	BAT47 ◆
(a)	40	350	0.4	10	◇	BAT48 ◆
(b)	80	1000	0.42	100	◇	BAT49 ◆
(f)	40	30	0.41	1	●	BAT81 ◆
(f)	60	30	0.41	1	●	BAT83 ◆
(f)	30	200	0.4	10	●	BAT85 ◆
(a)	60	300	1.0	200	●	BAV10
(a)	120	250	1.2	200	◇	BAV19
(a)	200	250	1.2	200	◇	BAV20
(a)	250	250	1.2	200	◇	BAV21
(g)	35	50	1.0	10	─	BAV45
(a)	75	100	1.0	100	●	BAW62
(a)	50	75	1.0	20	●	BAX1301
(a)	150	200	1.3	100	○	BAX16
(a)	150	200	1.3	100	○	BAX1601
(a)	150	200	1.3	100	○	BAX16ES
(a)	100	200	1.0	100	○	BAY72
(b)	100	200	0.8	10	─	BAY73
(b)	20	1000	0.55	1000	○	BYV10-20 ◆
(b)	30	1000	0.55	1000	○	BYV10-30 ◆
(b)	40	1000	0.55	1000	○	BYV10-40 ◆
(k)	60	1000	0.7	1000	○	BYV10-60 ◆
(k)	35	10	1.5	5	─	JPAD100
(k)	35	10	1.5	5	□	JPAD50
(j)	35	50	1.5	5	□	PAD100
(h)	45	50	1.5	5	□	PAD5
(e)	30	110	0.54	130	●	OA47 ■
(e)	30	10	2.0	30	○	OA90 ■
(e)	115	50	2.1	30	○	OA91 ■
(e)	115	50	1.85	30	○	OA95 ■
(a)	150	80	1.15	30	○	OA20201
(e)	400VRW	400	1.0	400	○	ZS104
(a)	100	75	1.0	10	●	1N914
(a)	100	75	1.0	10	●	1N916
(a)	125	200	0.8	10	□	1N3595
(a)	75	150	1.0	10	○	1N4148
(a)	75	200	1.0	10	●	1N4148-NSC
(a)	75	75	1.0	10	●	1N4149
(a)	50	200	0.74	20	●	1N4150
(a)	75	200	1.0	100	●	1N4446
(a)	75	150	1.0	100	●	1N4448
(a)	75	200	1.0	100	●	1N4448-NSC
(d)	40	75	1.0	10	○	1S44
(d)	50	200	1.2	200	○	1S920
(d)	100	200	1.2	200	○	1S921
(d)	150	200	1.2	200	○	1S922
(d)	200	200	1.2	200	○	1S923

■ Germanium   ◆ Silicon schottky barrier

Appendix F: Power diodes

Power diodes

Package (Not relative size)	IF (AV) Max Mean F'ward Current	VRRM								
		50—100	200	300	400	600	800	1000	1200	1600
L = 4.6 D = 3.8 GLASS	1A				1N5060					
L = 5 D = 2.7 ... PLASTIC	1A	1N4001 (50V) 1N4002 (100V)	1N4003		1N4004	1N4005	1N4006	1N4007		
		1N4001TR■ (50V) 1N4002TR■ (100V)	1N4003TR■		1N4004TR■	1N4005TR■	1N4006TR■	1N4007TR■		
		1N4001GP◆ (50V) 1N4002GP◆ (100V)	1N4003GP◆		1N4004GP◆	1N4005GP◆	1N4006GP◆	1N4007GP◆		
		1N4002GP◆ (100V)								
L = 6.35 D = 6.35 GLASS	3A		1N5624			1N5626				
L = 8.9 D = 3.7 PLASTIC	3A	30S1 (100V)	30S2		30S4	30S6	30S8	30S10		
L = 9.65 D = 5.3 PLASTIC	3A	1N5401 (100V)	MR502 1N5402		MR504 1N5404	1N5406		1N5408		
L = 9.1 D = 9.1 PLASTIC	6A	GI750 (50V) GI751 (100V)	GI752			GI756				
L = 9.5 D = 6.35 PLASTIC	6A	60S1 DIODE (100V)	60S2		60S4	60S6	60S8	60S10		
TO-220	6.5A			BY249-300		BY249-600				

Appendix F: Continued

Power diodes continued

Current	Package	100 V	200 V	300 V	400 V	600 V	800 V	1000 V	1200 V	1600 V
10A										
12A			12F20'' 12FR20'					12F100'		
15A				BYX99-300''					BYX99-1200''	
16A	10-32 UNF 2A	M16-100'' M16-100R'	M16-200'' M16-200R'		M16-400'' M16-400R'		M16-800'' M16-800R'		M16-1200'' M16-1200R'	
16A			16F20'' 16FR20''				16F80'' 16FR80''	16F100' 16FR100'	16F120'' 16FR120''	
30A	METRIC M5			BYX96-300'' BYX96-300R''		BYX96-600'' BYX96-600R''			BYX96-1200'' BYX96-1200R''	BYX96-1600'' BYX96-1600R''
40A (A)		M41-100'' M41-100R'	M41-200'' M41-200R'			M41-600'' M41-600R'				
40A (A)		40HF10'' 40HFR10'	40HF20'' 40HFR20'		40HF40' 40HFR40'	40HF60' 40HFR60'	40HF80' 40HFR80'	40HF100'' 40HFR100''	40HF120'' 40HFR120''	
47A (B)						BYX97-600''' BYX97-600R''			BYX97-1200'' BYX97-1200R''	BYX97-1600'' BYX97-1600R''
70A (A)	(A) ¼-28 UNF 2A	M71-100'' M71-100R'	M71-200'' M71-200R'		M71-400'' M71-400R'	M71-600'' M71-600R'	M71-800'' M71-800R'			
70A (A)	(B) METRIC M6	70HF10'' 70HFR10'	70HF20'' 70HFR20'		70HF40'' 70HFR40'	70HF60'' 70HFR60'	70HF80'' 70HFR80'	70HF100'' 70HFR100''	70HF120'' 70HFR120''	
150A (A)		45L10'' 45LR10'	45L20'' 45LR20'		45L40'' 45LR40'	45L60'' 45LR60'	45L80'' 45L80''	45L100'' 45L100''	45L120'' 45LR120''	
250A (B)	(A) ¼-20 UNF 2A (B) ⅜-16 UNF 2A	70U10'' 70UR10'	70U20'' 70UR20'		70U40'' 70U40'				70U120'' 70UR120''	

Important — Forward current ratings quoted on stud mounting devices are maximum rating. Manufacturer's data should always be consulted as in some cases devices have to be forced air cooled to obtain the maximum ratings quoted.

'' Denotes Stud Cathode
' Denotes stud anode
■ Denotes Bandoliered

♦ Glass passivated hermetically sealed construction, with proven reliability equal to MIL-S-19500

Epoxy-potted bridge rectifiers

Voltage	Current	Device No.
200	2	KBPC 102
400	2	KBPC 104
600	2	KBPC 106
800	2	KBPC 108
200	4	KBU 4D
800	4	KBU 4K
200	6	KBPC 802
800	6	KBPC 808
200	12	SKB 25/02
800	12	SKB 25/08
1200	12	SKB 25/12
50	25	KBPC 25005
200	25	KBPC 2502
600	25	KBPC 2506
200	35	KBPC 3502
600	35	KBPC 3506

Notes:
1. The bridge assembly should be mounted on a heat sink.
2. Current ratings are for resistive loads. When the rectifier is
   used on a battery or capacitive load the current rating
   should be multiplied by 0.8.

Appendix G: Zener diodes

Zener diodes

L = 4.5, D = 2.0   BZX55 Series — glass DO35
L = 4.25, D = 1.85   BZY79 Series — glass DO35
BZX70 Series — plastic
L = 12.5, D = 6.5
Rounded end indicates cathode

L = 4.8, D = 2.6   BZX85 Series — glass DO41
L = 5.2, D = 2.7   BZX85 Series — glass DO41
Axial lead types: Coloured band indicates cathode

L = 8.9, D = 3.7
1N5000 Series — plastic

L = 4.57, D = 3.81 BZT03 Series — glass SOD-57
L = 5.0, D = 4.5 BZW03 Series — glass SOD-64

BZY93 Series — 10/32 UNF 2A stud

BZY91 Series ¼ × 28 UNF Stud

These types are available as normal (stud cathode). Add suffix 'R' to Order Code if reverse polarity is required.

Mttr / Nominal Zener Voltage	Philips	SGS-Thomson	SGS-Thomson	Philips	Philips	Philips	—	Philips	Philips	Philips
WATTAGE (All ± 5% Voltage Tolerance)	400mW	500mW	1.3W	1.3W	2.5W	3W	5W	6W	20W	75W
2.4V	BZX79C2V4	BZX55C2V4	BZX85C2V4							
2.7V	BZX79C2V7	BZX55C2V7	BZX85C2V7							
3V	BZX79C3V0	BZX55C3V0	BZX85C3V0							
3.3V	BZX79C3V3	BZX55C3V3	BZX85C3V3				1N5333B			
3.6V	BZX79C3V6	BZX55C3V6	BZX85C3V6				1N5334B			
3.9V	BZX79C3V9	BZX55C3V9	BZX85C3V9				1N5335B			
4.3V	BZX79C4V3	BZX55C4V3	BZX85C4V3				1N5336B			
4.7V	BZX79C4V7	BZX55C4V7	BZX85C4V7				1N5337B			
5.1V	BZX79C5V1	BZX55C5V1	BZX85C5V1	BZX85C5V1			1N5338B			
5.6V	BZX79C5V6	BZX55C5V6	BZX85C5V6	BZX85C5V6			1N5339B			
6.2V	BZX79C6V2	BZX55C6V2	BZX85C6V2	BZX85C6V2			1N5341B			
6.8V	BZX79C6V8	BZX55C6V8	BZX85C6V8	BZX85C6V8			1N5342B			
7.5V	BZX79C7V5	BZX55C7V5	BZX85C7V5	BZX85C7V5	BZX70C7V5	BZT03C7V5	1N5343B		BZY93C7V5#	BZY91C7V5
8.2V	BZX79C8V2	BZX55C8V2	BZX85C8V2	BZX85C8V2	BZX70C8V2	BZT03C8V2	1N5344B		BZY93C8V2#	
9.1V	BZX79C9V1	BZX55C9V1	BZX85C9V1	BZX85C9V1	BZX70C9V1	BZT03C9V1	1N5346B		BZY93C9V1#	
10V	BZX79C10	BZX55C10	BZX85C10	BZX85C10	BZX70C10	BZT03C10	1N5347B	BZW03-C10 NEW	BZY93C10#	BZY91C10
11V	BZX79C11	BZX55C11	BZX85C11	BZX85C11	BZX70C11	BZT03C11	1N5348B		BZY93C11	
12V	BZX79C12	BZX55C12	BZX85C12	BZX85C12	BZX70C12	BZT03C12	1N5349B	BZW03-C12 NEW	BZY93C12#	BZY91C12
13V	BZX79C13	BZX55C13	BZX85C13	BZX85C13	BZX70C13	BZT03C13	1N5350B		BZY93C13#	
15V	BZX79C15	BZX55C15	BZX85C15	BZX85C15	BZX70C15	BZT03C15	1N5352B		BZY93C15#	BZY91C15#

Zener diodes continued

	BZX79C	BZX55C	BZX85C	BZY85C	BZX70C	BZT03C	1N	BZW03	BZY93C	BZY91C
16V	BZX79C16	BZX55C16	BZX85C16	BZY85C16	BZX70C16	BZT03C16	1N5353B		BZY93C16#	
18V	BZX79C18	BZX55C18	BZX85C18	BZY85C18	BZX70C18	BZT03C18	1N5355B		BZY93C18#	BZY91C18#
20V	BZX79C20	BZX55C20	BZX85C20	BZV85C20	BZX70C20	BZT03C20	1N5357B		BZY93C20#	BZY91C20
22V	BZX79C22	BZX55C22	BZX85C22	BZV85C22	BZX70C22	BZT03C22	1N5358B		BZY93C22	
24V	BZX79C24	BZX55C24	BZX85C24	BZV85C24	BZX70C24	BZT03C24	1N5359B	BZW03-C24 NEW	BZY93C24#	BZY91C24
27V	BZX79C27	BZX55C27	BZX85C27	BZV85C27	BZX70C27	BZT03C27	1N5361B	BZW03-C27 NEW	BZY93C27#	BZY91C27
30V	BZX79C30	BZX55C30	BZX85C30	BZV85C30	BZX70C30	BZT03C30	1N5363B		BZY93C30#	BZY91C30
33V	BZX79C33	BZX55C33	BZX85C33	BZV85C33	BZX70C33	BZT03C33	1N5364B		BZY93C33#	BZY91C33
36V	BZX79C36	BZX55C36	BZX85C36	BZV85C36	BZX70C36	BZT03C36	1N5365B	BZW03-C36 NEW	BZY93C36	BZY91C36
39V	BZX79C39	BZX55C39	BZX85C39	BZV85C39	BZX70C39	BZT03C39	1N5366B		BZY93C39#	
43V	BZX79C43	BZX55C43	BZX85C43	BZV85C43	BZX70C43	BZT03C43	1N5367B		BZY93C43	BZY91C43
47V	BZX79C47	BZX55C47	BZX85C47	BZV85C47	BZX70C47	BZT03C47	1N5368B	BZW03-C47 NEW	BZY93C47	BZY91C47
51V	BZX79C51	BZX55C51	BZX85C51	BZV85C51	BZX70C51	BZT03C51	1N5369B	BZW03-C51 NEW	BZY93C51	BZY91C51
56V	BZX79C56	BZX55C56	BZX85C56	BZV85C56	BZX70C56	BZT03C56	1N5370B		BZY93C56	
62V	BZX79C62	BZX55C62	BZX85C62	BZV85C62	BZX70C62	BZT03C62	1N5372B		BZY93C62	BZY91C62
68V	BZX79C68	BZX55C68	BZX85C68	BZV85C68	BZX70C68	BZT03C68	1N5373B		BZY93C68	BZY91C68
75V	BZX79C75	BZX55C75	BZX85C75	BZV85C75	BZX70C75	BZT03C75	1N5374B	BZW03-C75 NEW	BZY93C75#	BZY91C75
82V		BZX55C82	BZX85C82			BZT03C82	1N5375B	BZW03-C82 NEW		
91V		BZX55C91	BZX85C91			BZT03C91	1N5377B			
100V		BZX55C100	BZX85C100			BZT03C100	1N5378B			
110V		BZX55C110	BZX55C110			BZT03C110	1N5379B			
120V		BZX55C120	BZX85C120			BZT03C120	1N5380B			
130V		BZX55C130	BZX85C130			BZT03C130	1N5381B			
150V		BZX55C150	BZX85C150			BZT03C150	1N5383B			
160V		BZX55C160	BZX85C160			BZT03C160	1N5384B			
180V		BZX55C180	BZX85C180			BZT03C180	1N5386B			
200V		BZX55C200	BZX85C200			BZT03C200	1N5388B			
220V						BZT03C220				
240V						BZT03C240				
270V						BZT03C270				

Appendix H: Transistors

Transistor pin connections

TO18, TO5, TO39
TO205

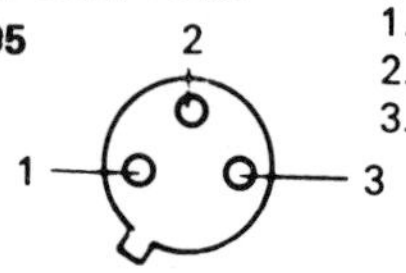

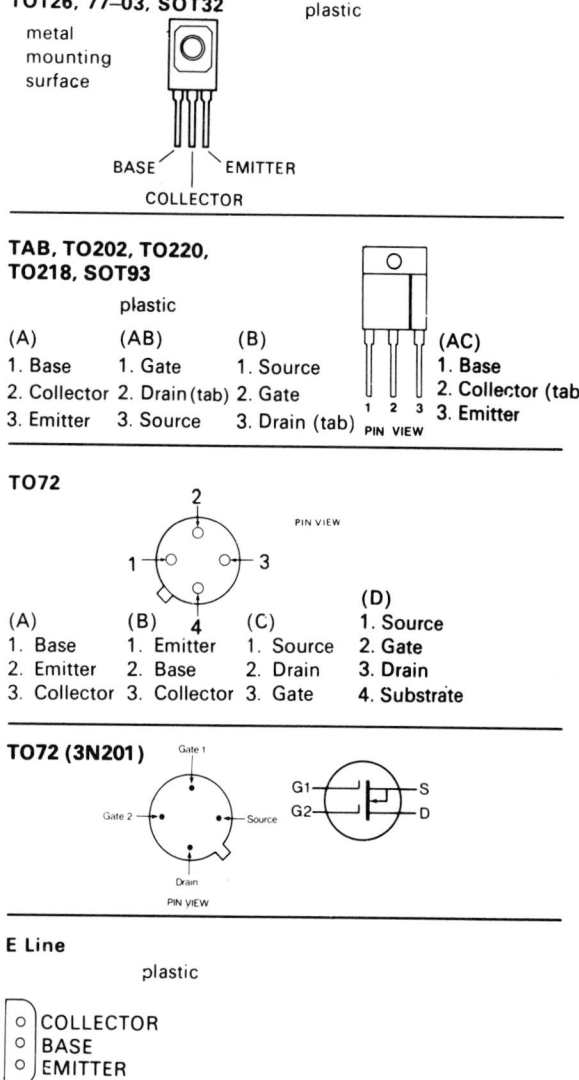

(A)	(B)	(C)
1. Emitter	1. Source	1. Source
2. Base	2. Drain	2. Gate
3. Collector	3. Gate	3. Drain
	(D) 1. Drain	
	2. Source	
	3. Gate	

TO3, SO55, TO204

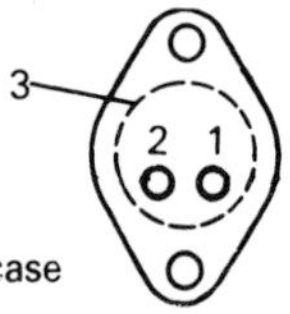

(A)	(B)	(C)
1. Base	1. Gate	1. Gate
2. Emitter	2. Drain	2. Source
3. Collector/case	3. Source/case	3. Drain/case

TO1

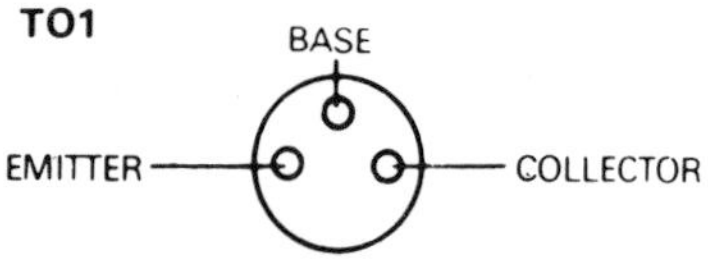

BASE
EMITTER —— COLLECTOR

SOT103

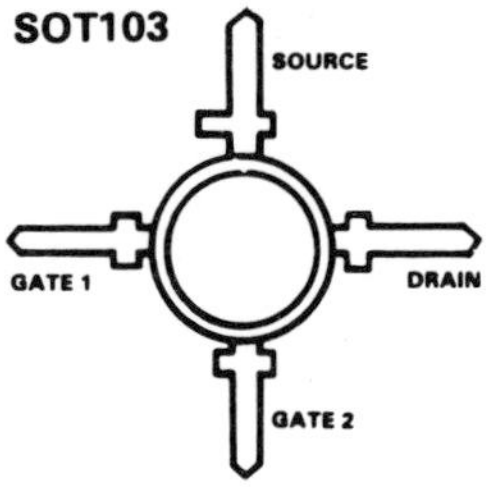

SOURCE
GATE 1   DRAIN
GATE 2

TO92
TO237

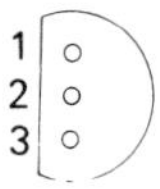

plastic

(A)	(B)	(C)	(D)
1. Base	1. Emitter	1. Collector	1. Drain
2. Collector	2. Base	2. Base	2. Gate
3. Emitter	3. Collector	3. Emitter	3. Source

(E)	(F)	(G)	(H)
1. Gate	1. Gate	1. Source	1. Source
2. Source	2. Drain	2. Gate	2. Drain
3. Drain	3. Source	3. Drain	3. Gate

TO126, 77–03, SOT32   plastic

metal mounting surface

BASE   EMITTER
COLLECTOR

TAB, TO202, TO220,
TO218, SOT93

plastic

(A)	(AB)	(B)	(AC)
1. Base	1. Gate	1. Source	1. Base
2. Collector	2. Drain (tab)	2. Gate	2. Collector (tab)
3. Emitter	3. Source	3. Drain (tab)	3. Emitter

1 2 3
PIN VIEW

TO72

PIN VIEW

(A)	(B)	(C)	(D)
1. Base	1. Emitter	1. Source	1. Source
2. Emitter	2. Base	2. Drain	2. Gate
3. Collector	3. Collector	3. Gate	3. Drain
			4. Substrate

TO72 (3N201)

Gate 1
Gate 2   Source
Drain
PIN VIEW

G1
G2
S
D

E Line   plastic

COLLECTOR
BASE
EMITTER

type	material	case	application	P_T	I_C	V_{CEO}	V_{CBO}	h_{FE}	f_T (typ)
AC127	NPN Ge	TO1	Audio output	340 mW	500 mA	12 V	32 V	50	2·5 MHz
AC128	PNP Ge	TO1	Audio output	700 mW	−1 A	−16 V	−32 V	60–175	1·5 MHz
AD149	PNP Ge	TO3(A)	Audio output	*22·5 W at 50 °C	−3·5 A	−50 V	−50 V	30–100	0·5 MHz
AD161 } Pair	NPN Ge } SO55(A)		Audio matched pair	*4 W at 72 °C	3 A	20 V	32 V	50–300	3 MHz
AD162 }	PNP Ge }			*6 W at 63 °C	−3 A	−20 V	−32 V	50–300	1·5 MHz
AF127	PNP Ge	TO72(A)	I.F. Applications	60 mW	−10 mA	−20 V	−20 V	150 ◆	75 MHz
BC107	NPN Si	TO18	Audio driver stages (complement BC177)	360 mW	100 mA	45 V	50 V	110–450	250 MHz
BC108	NPN Si	TO18	General purpose (complement BC178)	360 mW	100 mA	20 V	30 V	110–800	250 MHz
BC109	NPN Si	TO18	Low noise audio (complement BC179)	360 mW	100 mA	20 V	30 V	200–800	250 MHz
BC142	NPN Si	TO39	Audio driver	800 mW	800 mA	60 V	80 V	20 (min.)	80 MHz
BC143	PNP Si	TO39	Audio driver	800 mW	−800 mA	−60 V	−60 V	25 (min.)	160 MHz
BC177	PNP Si	TO18	Audio driver stages (complement BC107)	300 mW	−100 mA	−45 V	−50 V	125–500	200 MHz
BC178	PNP Si	TO18	General purpose (complement BC108)	300 mW	−100 mA	−25 V	−30 V	125–500	200 MHz
BC 179	PNP Si	TO18	Low Noise Audio (complement BC109)	300 mW	−100 mA	−20 V	−25 V	240–500	200 MHz
BC182L	NPN Si	TO92(A)	General purpose	300 mW	200 mA	50 V	60 V	100–480	150 MHz
BC183L	NPN Si	TO92(A)	General purpose (complement BC213L)	300 mW	200 mA	30 V	45 V	100–850	280 MHz
BC184L	NPN Si	TO92(A)	General purpose	300 mW	200 mA	30 V	45 V	250 (min.)	150 MHz
BC212L	PNP Si	TO92(A)	General purpose	300 mW	−200 mA	−50 V	−60 V	60–300	200 MHz
BC213L	PNP Si	TO92(A)	General purpose (complement BC183L)	300 mW	−200 mA	−30 V	−45 V	80–400	350 MHz
BC214L	PNP Si	TO92(A)	General purpose	300 mW	−200 mA	−30 V	−45 V	140–600	200 MHz
BC237B	NPN Si	TO92(B)	Amplifier	350 mW	100 mA	45 V	—	120–800	100 MHz
BC307B	PNP Si	TO92(B)	Amplifier	350 mW	100 mA	45 V	50 V	120–800	280 MHz
BC327	PNP Si	TO92(B)	Driver	625 mW	−500 mA	−45 V	−50 V	100–600	260 MHz
BC337	NPN Si	TO92(B)	Audio driver	625 mW at 45 °C	500 mA	45 V	50 V	100–600	200 MHz
BC441	NPN Si	TO39	General purpose (complement BC461)	1 W	2 A peak	60 V	75 V	40–250	50 MHz (min.)
BC461	PNP Si	TO39	General purpose (complement BC441)	1 W	−2 A peak	−60 V	−75 V	40–250	50 MHz
BC477	PNP Si	TO18	Audio driver stages	360 mW	−150 mA	−80 V	−80 V	110–950	150 MHz
BC478	PNP Si	TO18	General purpose	360 mW	−150 mA	−40 V	−40 V	110–800	150 MHz
BC479	PNP Si	TO18	Low noise audio amp.	360 mW	−150 mA	−40 V	−40 V	110–800	150 MHz
BCY70	PNP Si	TO18	General purpose	360 mW	−200 mA	−40 V	−50 V	150	200 MHz
BCY71	PNP Si	TO18	Low noise general purpose	360 mW	−200 mA	−45 V	−45 V	100–400	200 MHz
BD131	NPN Si	TO126^m	General purpose – medium power	*15 W at 60 °C	3 A	45 V	70 V	20 (min.)	60 MHz
BD132	PNP Si	TO126^m	General purpose – medium power	*15 W at 60 °C	−3 A	−45 V	−45 V	20 (min.)	60 MHz

Device	Type	Package	Application	Power	Current	Voltage	Voltage	hFE	Freq
BD131 } Pair	NPN Si } TO126^m		Audio matched pair	*15 W at 60 °C	3 A	45 V	70 V	20 (min.)	60 MHz
BD132 }	PNP Si }			*15 W at 60 °C	−3 A	−45 V	−45 V	20 (min.)	60 MHz
BD135	NPN Si	SOT32^m	Audio driver	*12·5 W at 25 °C	1·5 A	45 V	45 V	40–250	50 MHz
BD136	PNP Si	SOT32^m	Audio driver	*12·5 W at 25 °C	−1·5 A	−45 V	−45 V	40–250	75 MHz
BD437	NPN Si } TO126^m		Power switching complementary	*36 W at 25 °C	4 A	45 V	45 V	40	3 MHz
BD438	PNP Si }								
BCY70	PNP Si	TO18	General purpose	360 mW	−200 mA	−40 V	−50 V	150	200 MHz
BCY71	PNP Si	TO18	Low noise general purpose	360 mW	−200 mA	−45 V	−45 V	100–400	200 MHz
BD131	NPN Si	TO126^m	General purpose – medium power	*15 W at 60 °C	3 A	45 V	70 V	20 (min.)	60 MHz
BD132	PNP Si	TO126^m	General purpose – medium power	*15 W at 60 °C	−3 A	−45 V	−45 V	20 (min.)	60 MHz
BD131 } Pair	NPN Si } TO126^m		Audio matched pair	*15 W at 60 °C	3 A	45 V	70 V	20 (min.)	60 MHz
BD132 }	PNP Si }			*15 W at 60 °C	−3 A	−45 V	−45 V	20 (min.)	60 MHz
BD135	NPN Si	SOT32^m	Audio driver	*12·5 W at 25 °C	1·5 A	45 V	45 V	40–250	50 MHz
BD136	PNP Si	SOT32^m	Audio driver	*12·5 W at 25 °C	−1·5 A	−45 V	−45 V	40–250	75 MHz
BD437	NPN Si } TO126^m		Power switching complementary	*36 W at 25 °C	4 A	45 V	45 V	40	3 MHz
BD438	PNP Si }								
BD679	NPN Si } TO126^m		Audio complementary Darlington	*40 W at 25 °C	6 A	80 V	100 V	2200	60 kHz
BD680	PNP Si }			*40 W at 25 °C	6 A	80 V	80 V	2200	60 kHz
BDX33C	NPN Si } TO220(A)^m (A)		Power switching Darlington	*70 W at 25 °C	10 A	100 V	100 V	750 (min.)	—
BDX34C	PNP Si }								
BF259	NPN Si	TO39	High-voltage video amplifier	800 mW	100 mA	300 V	300 V	25	90 MHz
BF337	NPN Si	TO39	Video amplifier	*3 W at 125 °C	100 mA	200 V	250 V	20 (min.)	80 MHz

BFY50	NPN Si	TO39	High voltage general purpose	800 mW	1 A	35 V	80 V	30	60 MHz
BFY51	NPN Si	TO39	General purpose	800 mW	1 A	30 V	60 V	40	50 MHz
BFY52	NPN Si	TO39	General purpose	800 mW	1 A	20 V	40 V	60	50 MHz
TIP31A	NPN Si	TO220^m(A)	Plastic medium power complementary	*40 W at 25 °C	3 A	60 V	60 V	10–60	8 MHz
TIP32A	PNP Si							10–40	
TIP31C	NPN Si	TO220(A)	Plastic medium power complementary	*40 W at 25 °C	3 A	100 V	100 V	10–50	8 MHz
TIP32C	PNP Si								
TIP33A	NPN Si	TAB^m(A)	Audio output complementary	*80 W at 25 °C	10 A	60 V	60 V	20–100	3 MHz
TIP34A	PNP Si								
2N2905	PNP Si	TO5(A)	Switching	600 mW	−600 mA	−40 V	−60 V	100–300	200 MHz (min.)
2N2905A	PNP Si	TO5(A)	Switching	600 mW	−600 mA	−60 V	−60 V	100–300	200 MHz (min.)
2N2907A	PNP Si	TO18(A)	Switching	400 mW	−600 mA	−60 V	−60 V	100–300	200 MHz (min.)
2N3019	NPN Si	TO39(A)	General purpose	500 mW	1 A	80 V	140 V	90 min.	100 MHz
2N3053	NPN Si	TO39(A)	General purpose	800 mW	1 A	40 V	60 V	50–250	100 MHz
2N3055E	NPN Si	TO3(A)	High power epitaxial (complement MJ2955)	*115 W at 25 °C	15 A	60 V	100 V	20–70	2·5 MHz
2N3055H	NPN Si	TO3(A)	High power homotaxial (complement PNP3055)	*115 W at 25 °C	15 A	60 V	100 V	20–70	1 MHz
PNP3055	PNP Si	TO3(A)	High power (complement 2N3055H)	*115 W at 25 °C	−15 A	−60 V	−100 V	20–70	0·8 MHz
2N3440	NPN Si	TO39(A)	General purpose	1 W	1 A	250 V	300 V	40–160	15 MHz
2N3702	PNP Si	TO92(A)	General purpose	360 mW	−200 mA	−25 V	−40 V	60–300	100 MHz

Appendix I: Voltage regulators

Voltage regulators

Fixed voltage series regulators

Pin-out diagrams:

- 78 and 78S series (B C E) — I/P, COM, O/P   or   I/P, COM, O/P
- 79 series (B C E) — COM, O/P, I/P   or   COM, O/P, I/P
- 78L series — O/P, I/P, COM
- 791 series — COM, O/P, I/P
- TO3 — I/P, O/P, Base common

Description	Output Voltage	Current	Case	Stock No.	Suitable transformer Stock No.	Sec. Voltage	VA
78L05	5 V	100 mA	TO92	306-190	207-188	9(S)	6
RS309K (LM309K)	5 V	1.2 A	TO3	305-614	207-122	9(S)	20
7805	5 V	1.0 A	TO220	305-888	207-122	9(S)	20
78S05	5 V	2 A	TO220	633-026	207-239	9(S)	50
78H05	5 V	5 A	TO3	307-301			
78L12	12 V	100 mA	TO92	306-207	207-217	15(P)	6
7812	12 V	1.0 A	TO220	305-894	207-267	15(P)	50
78S12	12 V	2 A	TO220	633-032	207-267	15(P)	50
78H12	12 V	5 A	TO3	307-317	207-289	12(S)	100
78L15	15 V	100 mA	TO92	306-213	207-217	15(P)	6
7815	15 V	1.0 A	TO220	305-901	207-267	15(P)	50
78S15	15 V	2 A	TO220	633-048	207-267	15(P)	50
78L24	24 V	100 mA	TO92	306-229	207-201	24(S)	6
7824	24 V	1.0 A	TO220	305-917	207-251	24(S)	50
78S24	24 V	2 A	TO220	633-054	207-295	24(S)	100
79L05	-5 V	-100 mA	TO92	306-235	207-188	9(S)	6
7905	-5 V	-1.2 A	TO220	306-049	207-122	9(S)	20
79L12	-12 V	-100 mA	TO92	306-241	207-217	15(P)	6
7912	-12 V	-1.2 A	TO220	306-055	207-267	15(P)	50
79L15	-15 V	-100 mA	TO92	306-257	207-217	15(P)	6
7915	-15 V	-1.2 A	TO220	305-923	207-267	15(P)	50
79L24	24 V	-100 mA	TO92	306-263	207-201	24(S)	6
7924	-24 V	-1.0 A	TO220	306-184	207-251	24(S)	50

Source: RS Data Library

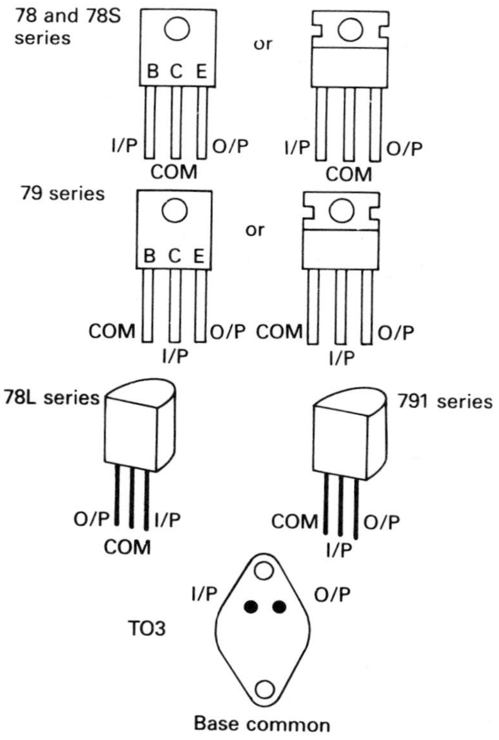

78H05 and 78H12 fixed hybrid regulators

Two fixed voltage hybrid regulators, housed in T03 style metal cases, capable of supplying output currents up to 5 amps. The internal circuitry limits the junction temperature to a safe value and provides automatic thermal overload protection.

Safe operating protection is also incorporated making the regulators virtually damage proof.

In order to achieve maximum performance the internal power dissipation must be kept below 50 W. Transformer and heatsink selections are dependent upon the exact application.

78H – Basic circuit, fixed voltage

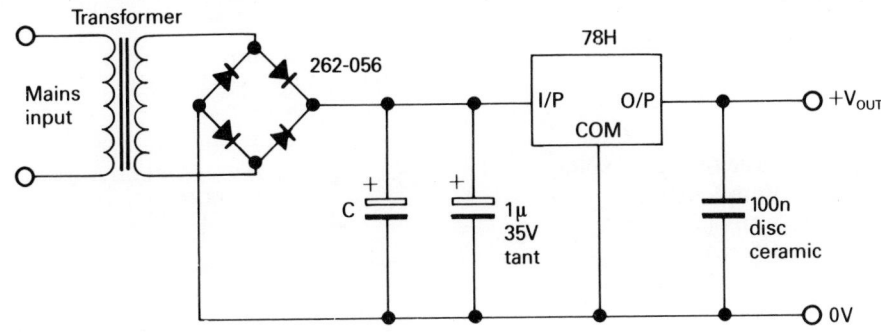

Regulator	Output voltage	Transformer	Heatsink	C
78H05	+ 5 V dc	207-239 (S)	401-807,1/1°c/w	15,000 μF 16 V
78H12	+ 12 V dc	207-289 (S)	401-403,2/1°c/w	22,000 μ 25 V

Source: RS Data Library

79 Series negative regulators

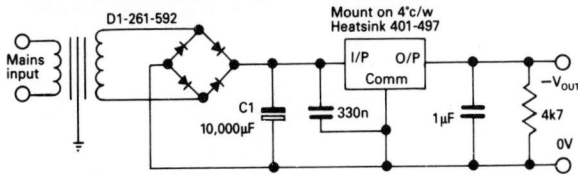

78 Series positive regulators

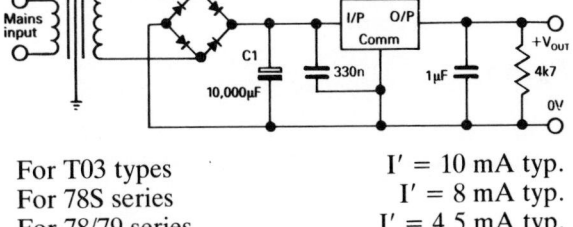

For T03 types	I′ = 10 mA typ.
For 78S series	I′ = 8 mA typ.
For 78/79 series	I′ = 4.5 mA typ.
For 78L/79L series	I′ = 3.5 mA typ.

Circuit gives constant current through load provided V_{out} does not exceed $V_{in} - (V_R + 2.5)$. [$V_{in} - (V_R + 3)$ for 78S series.] Select R to give designed constant current I_{out}

Constant current generator

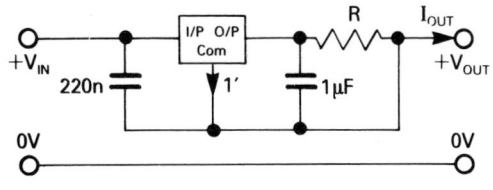

$$I_{out} = \frac{V_R}{R} + I'$$

Where V_R is the basic regulator voltage.

Source: RS Data Library

Adjustable voltage regulators

Increasing basic regulator voltage

The input voltage V_{in} should be derived from a suitable transformer, rectifier and smoothing capacitor circuit. Note V_{in} must be greater (within maximum ratings) than $V_{out} + 2.5$ V.

Figure 5 – gives higher output voltage than basic circuit but with reduced regulation.

$$V_{out} = V_R \left(1 + \frac{R_2}{R_1}\right) + I'R_2$$

$$I_{R1} \geqslant 5 \times I'$$

where V_R = basic regulator voltage
$I' = 10$ mA (T03)
$= 8$ mA (78S series)
$= 4.5$ mA (78/79 series)
$= 3.5$ mA (78L/79L series)

Figure 6 – gives better regulation than Figure 5.
$$V_{out} = V_R + V_1 + 0.6$$

where $V_1 = \dfrac{R_2 V_{out}}{R_1 + R_2}$

and $\dfrac{R_1}{R_2} = \dfrac{V_R + 0.6}{V_{out} - (V_R + 0.6)}$

e.g. For 9 V output with 5 V regulator
R1 = 5 k6   R2 = 3 k3

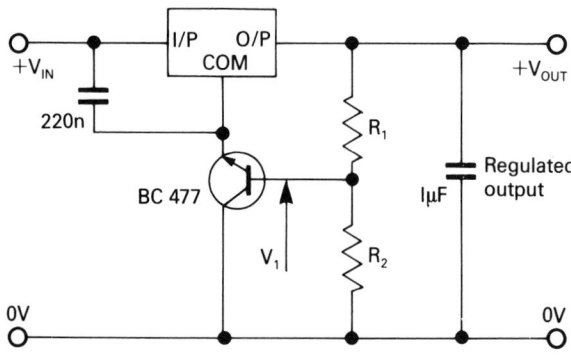

Figure 6

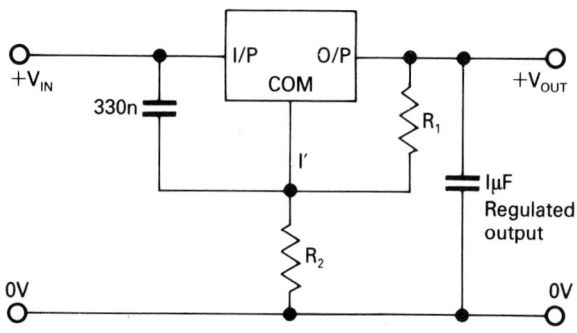

Source: RS Data Library

Figure 5

Appendix J: Power control

Power control, thyristor ratings

Device No.	Max. volts	Max. current	Gate I	Gate V
C203YY	60 V	0.8 A	0.2 mA	0.8 V
BTX18-400	500 V	1.0 A	5 mA	2 V
BT106	700 V	1.0 A	50 mA	3.5 V
C106	400 V	2.55 A	0.2 mA	0.8 V
2N4443	400 V	5.1 A	30 mA	1.5 V
2N4444	600 V	5.1 A	30 mA	1.5 V
BT152-600	600 V	13 A	32 mA	1 V
BTY79-400R	400 V	6.4 A	30 mA	3 V
BTY79-800R	800 V	6.4 A	30 mA	3 V
N018RH05	500 V	21 A	100 mA	3 V
N018RH08	800 V	21 A	100 mA	3 V
N018RH12	1200 V	21 A	100 mA	3 V
N029RH05	500 V	30 A	100 mA	3 V
N029RH08	800 V	30 A	100 mA	3 V
N029RH12	1200 V	30 A	100 mA	3 V
N044RH05	500 V	45 A	100 mA	3 V
N044RH08	800 V	45 A	100 mA	3 V
N044RH12	1200 V	45 A	100 mA	3 V
N060RH06	600 V	63 A	100 mA	3 V
N060RH08	800 V	63 A	100 mA	3 V
N060RH12	1200 V	63 A	100 mA	3 V
N086RH06	600 V	85 A	150 mA	3 V
N086RH08	800 V	85 A	150 mA	3 V
N086RH12	1200 V	85 A	150 mA	3 V
N105RH06	600 V	110 A	150 mA	3 V
N105RH08	800 V	110 A	150 mA	3 V
N105RH12	1200 V	110 A	150 mA	3 V

Power control, triac ratings

Device No.	Case type	Max. volts	Max. current	Gate V
Z0105DA	T092	400 V	0.35 A	2.0 V
T1CP206D	T092	400 V	1.5 A	2.5 V
T1CP206M	T092	600 V	1.5 A	2.5 V
T1C206M	T0220AB	600 V	4.0 A	2.0 V
T1C216M	T0220AB	600 V	6.0 A	3.0 V
T1C225M	T0220AB	600 V	8.0 A	2.0 V
T1C226M	T0220AB	600 V	8.0 A	2.0 V
T1C236M	T0220AB	600 V	12.0 A	2.0 V
BT139	T0220AB	600 V	15.0 A	1.5 V
T1C246M	T0220AB	600 V	16.0 A	2.0 V

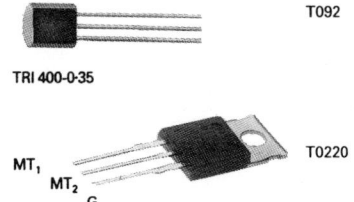

T092

TRI 400-0-35

MT₁
MT₂
G

T0220

Figure 6.25   Appearance of a triac

Diac trigger diodes

Device No.	Trigger volts	Max. current
BR100D0-14	32 V ± 4 V	2
133D0-7	32 V ± 4 V	1
D0201YR	32 V ± 4 V	1

Silicon bidirectional trigger diodes for use in triac firing circuits are glass encapsulated with the appearance of an axial lead small signal diode. (See Appendix E for photograph.)

Appendix K: Comparison of British and American logic gate symbols

Comparison of British and American logic gate symbols

Logic gate	American symbol	British symbol	Truth table
AND			A B OUT / 0 0 0 / 0 1 0 / 1 0 0 / 1 1 1
OR			A B OUT / 0 0 0 / 0 1 1 / 1 0 1 / 1 1 1
Exclusive - OR			A B OUT / 0 0 0 / 0 1 1 / 1 0 1 / 1 1 0
NOT			A OUT / 0 1 / 1 0
NOR			A B OUT / 0 0 1 / 0 1 0 / 1 0 0 / 1 1 0
NAND			A B OUT / 0 0 1 / 0 1 1 / 1 0 1 / 1 1 0

Appendix L: Integrated circuit logic gates

CMOS integrated circuit logic gates

4001B Quad 2 input NOR
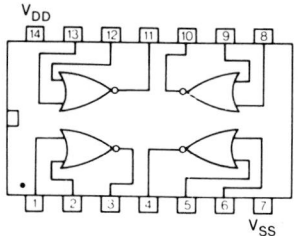

4002B Dual 4 input NOR
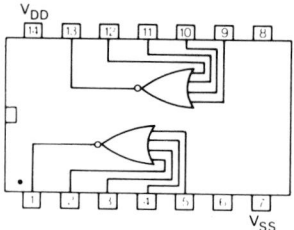

4023B Triple 3 input NAND
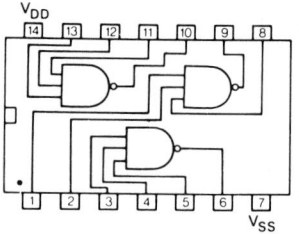

4011B Quad 2 input NAND
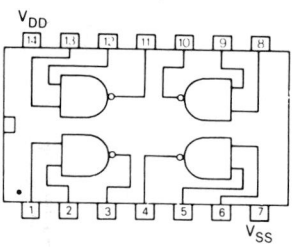

4012B Dual 4 input NAND
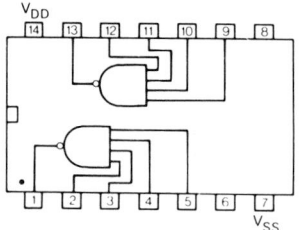

4013B Dual D type flip flop
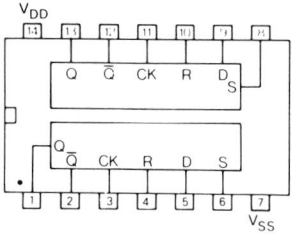

4025B Triple 3 input NOR
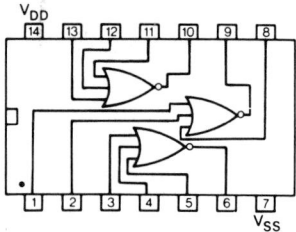

4027B Dual J.K. flip flop
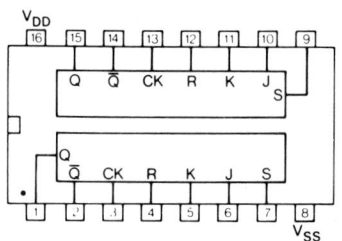

4049UB Hex inverter buffer
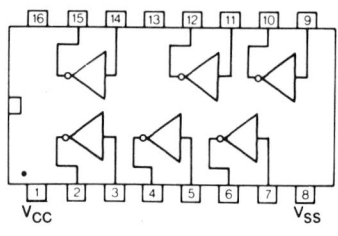

4069UB Hex inverter
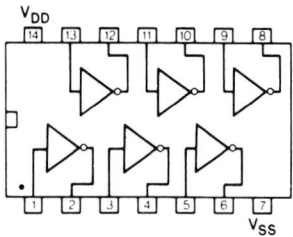

4071B Quad 2 input OR
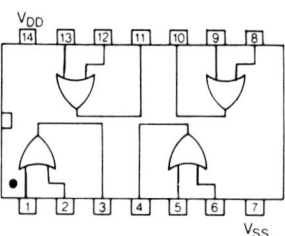

4072B Dual 4 input OR gate
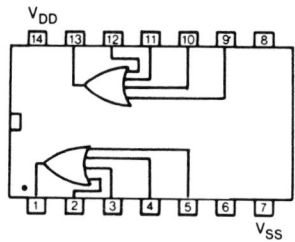

4073B Triple 3 input AND
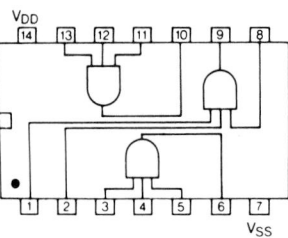

4075B Triple 3 input OR
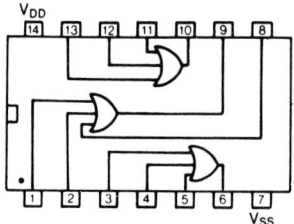

4081B Quad 2 input AND
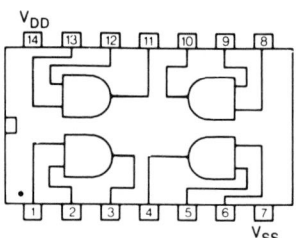

TTL integrated circuit logic gates

7400 Quadruple 2-input NAND gate

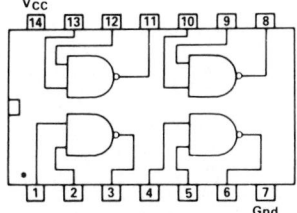

7402 Quadruple 2-input NOR gate

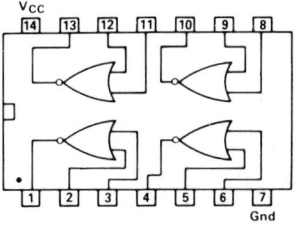

7404 Hex inverter

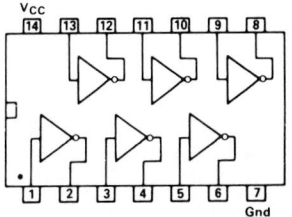

7408 Quadruple 2-input AND gate

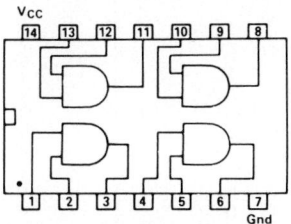

7410 Triple 3-input NAND gate

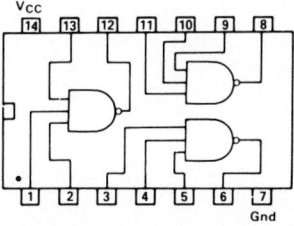

7411 Triple 3-input AND gate

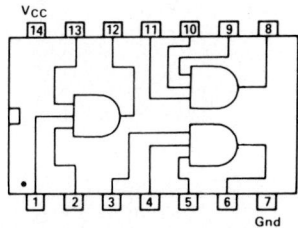

7414 Hex Schmitt Trigger

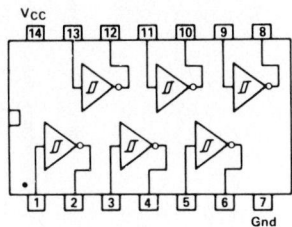

7420 Dual 4-input NAND gate

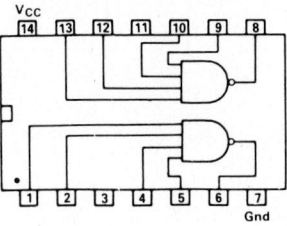

7421 Dual 4-input AND gate

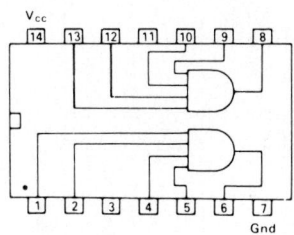

7427 Triple 3-input NOR gate

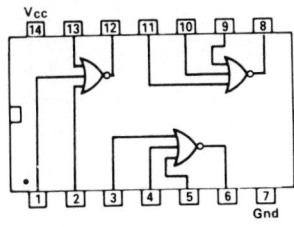

7432 Quadruple 2-input OR gate

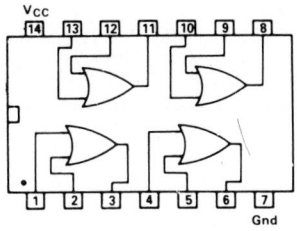

7474 Dual D-type edge-triggered Flip Flop

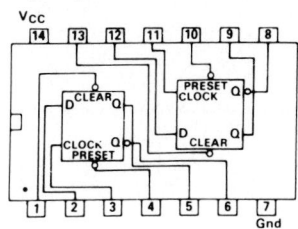

7476 Dual JK Flip Flop with set and clear

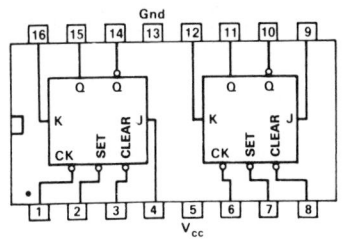

Appendix M: Thermocouple colour coding

Thermocouple colour coding

Generic and trade names	Colour Coding			Magnetic		Maximum useful temp. range	EMF (MV) over useful temp. range	Average sensitivity μV/°C	Environment (bare wire)
	Single	Overall T/C wire	Overall extension grade wire	Yes	No				
Copper	Blue	Brown	Blue		X	°F −328 to 662	−5.602 to 17.816	40.5	Mild oxidising, reducing. Vacuum or inert. Good where moisture is present
Constantan, Cupron, Advance	Red				X	°C −200 to 350			
Iron	White	Brown	Black	X		°F 32 to 1382	0 to 42.283	52.6	Reducing. Vacuum, inert. Limited use in oxidising at high temperatures. Not recommended for low temps
Constantan, Cupron, Advance	Red				X	°C 0 to 750			
Chromel, Tophel, T1 Thermokanthal KP	Purple	Brown	Purple		X	°F −328 to 1652	−8.824 to 68.783	67.9	Oxidising or inert. Limited use in vacuum or reducing
Constantan, Cupron, Advance	Red				X	°C −200 to 900			
Chromel, Tophel, T1 Thermokanthal KP	Yellow	Brown	Yellow	X		°F −328 to 2282	−5.973 to 50.633	38.8	Clean oxidising and inert. Limited use in vacuum or reducing
Alumel, Nial, T2 Thermokanthal KN	Red				X	°C −200 to 1250			
Platinum 10% Rhodium	Black		Green		X	°F 32 to 2642	0 to 14.973	10.6	Oxidising or inert atmos. Do not insert in metal tubes. Beware of contamination
Pure Platinum	Red				X	°C 0 to 1450			
Platinum 13% Rhodium	Black		Green		X	°F 32 to 2642	0 to 16.741	12.0	
Pure Platinum	Red				X	°C 0 to 1450			
Platinum 30% Rhodium	Grey		Grey		X	°F 32 to 3092	0 to 12.426	7.6	
Platinum 6% Rhodium	Red				X	°C 0 to 1700			
Tungsten 5% Rhenium	White/red trace		White/red trace		X	°F 32 to 4208	0 to 37.066	16.6	Vacuum, inert, hydrogen atmospheres. Beware of embrittlement
Tungsten 26% Rhenium	Red				X	°C 0 to 2320			
Tungsten	White/blue trace		White/Blue trace		X	°F 32 to 4208	0 to 38.414	16.0	
Tungsten 26% Rhenium	Red				X	°C 0 to 2320			
Tungsten 3% Rhenium	White/yellow trace		White/yellow trace		X	°F 32 to 4208	0 to 39.506	17.0	
Tungsten 25% Rhenium	Red				X	°C 0 to 2320			

Notes:

1. Standard ANSI colour coding is used on all insulated thermocouples wire and extension grade wire when the insulation permits. For some insulations a coloured tracer is used to distinguish the calibration.

2. Thermocouple and extension grade wires are specified by ANSI letter designations. Positive and negative legs are identified by the letter suffixes P and N respectively. Example: JP designates the positive leg (iron) of the Iron-Constantan pair.

Appendix N: Strain gauges

Strain gauges

General specification (all types)

Measurable strain	2 to 4% maximum
Thermal output 20 to 160°C	± 2 micro strain/°C*
160 to 180°C	± 5 micro strain/°C*
Gauge factor change with temperature	±0.015%/°C max
Gauge resistance	120 Ω
Gauge resistance tolerance	±0.5%
Fatigue life	>10^5 reversals @ 1000 micro strain*
Foil material	copper nickel alloy

*1 micro strain is equivalent to an extension of 0.0001%

**Specification
(standard polyester backed types)**

Temperature range	−30°C to +80°C
Gauge length	8 mm
Gauge width	2 mm
Gauge factor	2.1
Base length (single types)	13.0 mm
Base width (single types)	4.0 mm
Base diameter (rosettes)	21.0 mm

**Specification
(miniature polyimide backed type)**

Temperature range	−30°C to +180°C
Gauge length	2 mm to 5 mm
Gauge width	1.6 mm to 1.8 mm
Gauge factor	2.0 to 2.1
Base length (single types)	6.0 mm to 9.0 mm
Base width (single types)	2.5 mm to 3.5 mm
Base size (rosettes)	7.5 × 7.5 mm, 12 × 12 mm

GLOSSARY

Adder   In a computer it is a device which can form the sum of two or more numbers.

Address   Information which identifies a particular location in the memory of a computer.

Aerial   The part of a communication system from which energy is radiated or received.

ALGOL   A symbolic language used to program computers in mathematical and engineering applications.

Alignment   The adjustment of tuned circuits so that they respond in a desired way at a given frequency.

AM (amplitude modulation)   A method of sending a message on a radio or light wave by varying the amplitude of the wave in response to the frequency of the message.

Amplifier   A device whose output is a magnified function of its input.

Analogue signals   Signals that respond to or produce a continuous range of values rather than specific values.

Analogue-to-digital converter   A circuit designed to convert an analogue voltage into a binary code which can be read by a computer.

AND gate   A building block in digital logic circuits.

Astable   A circuit which can generate a continuous waveform with no trigger

Atom   The smallest particle of a chemical element that can exist alone or in combination with other atoms.

Attenuator   A network designed to reduce the amplitude of a wave without distortion.

Audio frequency   Any frequency at which the sound wave can normally be heard. The audio frequencies for most humans are those frequencies between 15 Hz and about 20 kHz.

Automation   Any device or system that takes the place of humans in carrying out repetitive and boring jobs.

Band pass filter   A filter that passes all frequencies between two specified frequencies.

Bandwidth   The range of frequencies amplified or the range of frequencies passed by a filter.

Basic (beginners all-purpose symbolic instruction code) An introductory high-level computer programming language.

Battery   A power source made up of a number of individual cells.

Bias   The current or voltage which is applied to part of a circuit to make the circuit function properly.

Binary number   A number which can have just two values, 1 and 0.

Bipolar transistor   A transistor that depends, for its operation, on both n-type and p-type semiconductors.

Bistable (also flip-flop)   A circuit which has two stable outputs which can act as memories for data fed into its input.

Bit   A unit of information content.

Boolean algebra   A branch of symbolic logic in which logical operations are indicated by operators such as AND, OR and NOT signs.

Bourdon tube   A pressure-measuring device made of a flexible tube formed into a C shape. Increasing pressure causes the C to straighten.

Buffer   An isolating circuit used to avoid a reaction between a driver and driven circuit.

BUS   One or more conductors used as a path for transmitting information from source to destination.

Byte   A sequence of adjacent binary bits, usually 8.

Carrier wave   A relatively high frequency wave which is suitable for transmission and modulation.

CATV (cable TV)   A distribution of TV programmes by means of cables laid underground.

Cell   A single source of electric potential.

Chip   A small piece of silicon on which a complex miniaturised circuit, called an integrated circuit, is formed by photographic and chemical processes.

Circuit   An electrical network in which there is at least one path which can be closed.

Closed loop control   A control system which modifies its own behaviour according to feedback information e.g. constant speed control of an electric motor.

Closed loop gain   The gain of an amplifier with feedback. Negative feedback reduces amplifier gain. Positive feedback increases amplifier gain.

CMOS (complementary metal oxide semiconductor logic)   A logic family used especially in portable equipment.

Coaxial cable   A cable formed from an inner and outer cylindrical conductor.

Code   A system of symbols which represents information in a form which is convenient for a computer.

Colour code   The values and tolerances of components such as resistors indicated by coloured bands.

Combinational logic   A digital circuit, e.g. a NAND gate, that produces an output based on the combination of 0s and 1s presented to its input.

Communications   The transmission of information by means of electromagnetic waves or by signals along conductors.

Comparator   An electronic device, e.g. one based on an operational amplifier, that produces an output when the voltages of two input signals are different.

Computer   A programmable device used for storing, retrieving and processing data.

CPU (central processor unit)   The principal operating and controlling part of a computer, also known as its microprocessor.

Critically damped   The degree of damping that provides the best compromise between the undamped response and the overdamped response.

CRO (cathode ray oscilloscope)   A test and measurement instrument for showing the patterns of electrical waveforms and for measuring their frequency and other characteristics.

Cutoff frequency   The 'corner frequency' of a filter. That frequency at which the signal level falls by 3 dB.

Data-handling   Automatic or semi-automatic equipment which can collect, receive, transmit and store numerical data.

Decade counter   A binary counter that counts up to a maximum count of ten before resetting to zero.

Decibel (dB)   A unit used for comparing the strengths of two signals, such as the intensity of sound and the voltage gain of an amplifier.

Decoder   A device that converts coded information, e.g. the binary code into a more readily understood code such as decimal.

Decoupling network   A network designed to prevent an interaction between two electric circuits. These usually consist of RL or RC filters.

Demodulator   A device for recovering information, e.g. music from a carrier wave.

DIAC   Four-layer breakover device used to extend the range of control in a TRIAC circuit.

Dielectric   The insulating layer between the conducting plates of a capacitor.

Difference amplifier   An operational amplifier circuit that finds the difference between two input voltages.

Differential pressure flowmeter   A device for measuring the flow of fluids. Depends on the drop in pressure created when a fluid flows past an obstruction or around a bend.

Differentiator   An amplifier that performs the calculus operation of differentiation.

Diffusion   The movement of electrons and/or holes from a region of high to low concentration.

Digital computer   A system that uses gates, flip-flops, counters etc., to process information in digital form.

Digital-to-analogue converter   A device that converts a digital signal into an equivalent analogue signal. DACs are widely used in computer systems for controlling the speed of motors, the brightness of lamps etc.

Digital voltmeter   A voltmeter which displays the measured value as numbers composed of digits.

Doping   The process of introducing minute amounts of material, the dopant, into a silicon crystal to produce n-type or p-type semiconductors in the making of transistors, integrated circuits and other devices.

Earth electrode   A conductor driven into the earth and used to maintain conductors connected to it at earth potential.

Electric charge   The quantity of electricity contained in or on a body, symbol Q, measured in coulombs, symbol c.

Electrolyte   A substance which produces a conducting medium when dissolved in a suitable solvent.

Electrolytic capacitor   A capacitor which is made from two metal plates separated by a very thin layer of aluminium oxide. Electrolytic capacitors offer a high capacitance in a small volume, but they are polarised and need connecting the right way round in a circuit.

Electromagnetic spectrum   The family of radiations which all travel at the speed of light through a vacuum and include light, infrared and ultraviolet radiation.

Electron   A small negatively charged particle which is one of the basic building blocks of all substances and forms a cloud round the nucleus of an atom.

Electronic ice   A system of reference junction compensation used in thermocouple circuits. This is an electronic means of creating the thermocouple reference junction.

Electronics   That branch of science and technology which is concerned with the study of the conduction of electricity in a vacuum and in semiconductors and with the application of devices using these phenomena.

E.M.F. (electromotive force)   The electrical force generated by a cell or battery that makes electrons move through a circuit connected across the terminals of the battery.

Encode   Any device that converts information into a form suitable for transmission by electronic means.

Fan-in   The number of logic gate outputs which can be connected to the input of another logic gate.

Fan-out   The number of logic gate inputs which may be driven from a logic gate output.

Farad (F)   The unit of electrical capacitance and equal to the charge stored in coulombs in a capacitor when the potential difference across its terminals is 1 V.

FAX (facsimile)   The process of scanning fixed graphic material so that the image is converted into an electrical signal which may be used to produce a recorded likeness of the original.

Feedback   The sending back to the input part of the output of a system in order to improve the performance of the system. There are two types of feedback, positive and negative.

Ferrite   One of a class of magnetic materials which have a very low eddy-current loss, used for high-frequency circuit transformers and computer memories.

Ferroxcube   A commercially available ferrite.

FET (field-effect transistor)   A unipolar transistor that depends for its operation on either n-type or p-type semiconductor material.

Fibre optics   The use of hair-thin transparent glass fibres to transmit information on a light beam that passes through the fibre by repeated internal reflections from the walls of the fibre.

Fidelity   The quality or precision or the reproduction of sound.

Filter   A circuit that passes only signals of a desired frequency or band of frequencies. May be high pass, low pass or band pass.

Flip-flop   A device having two stable states, logic 0 or logic 1, and two input terminals corresponding to these states. The device will remain in either state until caused to change to the other by the application of an appropriate signal. In digital electronics, the bistable multivibrator circuit has earned the name 'flip-flop'.

Float switch   Level-sensing limit switch, actuated by a float on the surface of a liquid.

Floppy disc   A flexible disc, usually 5.25 inches (133 mm) in diameter, made of plastic and coated with a magnetic film on which computer data can be stored and erased.

FM (frequency modulation)   A method of sending information by varying the frequency of a radio or light wave in response to the amplitude of the message being sent. For high-quality radio broadcasts, FM is preferable to AM since it is affected less by interference from electrical machinery and lightning.

Force   A directed effort that changes the motion of a body.

Forward bias   A voltage applied across a pn junction which causes electrons to flow across the junction.

Forward breakover voltage   The voltage between anode and cathode of an SCR at which forward bias conduction will begin.

Fourier analysis   A mathematical method of determining the harmonic component of a complex wave.

Gain   The ratio of increase in signal level between the input and output of an amplifier.

Gauge factor   The ratio of change in resistance to the change in length of a strain gauge. Approximately two for a bonded foil strain gauge.

Gigabyte (GB)   A quantity of computer data equal to one thousand million bytes.

Gigahertz (GHz)   A frequency equal to one thousand million hertz (10^9 Hz)

Half-wave rectifier   A diode, or circuit based on one or more diodes, which produces a direct current from alternating current by removing one half of the AC waveform.

Hardware   Any mechanical or electronic equipment that makes up a system.

Heat sink   A relatively large piece of metal that is placed in contact with a transistor or other component to help dissipate the heat generated within the component.

Henry (H)   The unit of electrical inductance.

Hertz (Hz)   The unit of frequency equal to the number of complete cycles per second of an alternating waveform.

High pass filter   A filter that passes all frequencies above a specified frequency.

Hole   A vacancy in the crystal structure of a semiconductor that is able to attract an electron. A p-type semiconductor contains an excess of holes.

Impedance (Z)   The resistance of a circuit to alternating current.

Impurity   An element such as boron that is added to silicon to produce a semiconductor with desirable electrical qualities.

Inductor   An electrical component, usually in the form of a coil of wire. Inductors are used as 'chokes' to reduce the possibly damaging effects of sudden surges of current, and in tuned circuits.

Information technology (IT)   The gathering, processing and circulation of information by combining the data-processing power of the computer with the message-sending capability of communications.

Input/output port   The electrical 'window' on most computer systems that allows the computer to send data to and receive data from an external device.

Instrumentation amplifier   A difference amplifier with very high input impedances at both inputs.

Insulator   A material, e.g. glass, that does not allow electricity to pass through it.

Integrated circuit (IC)   An often very complex electronic circuit which has resistors, transistors, capacitors and other components formed on a single silicon chip.

Integrator   An amplifier circuit that performs the calculus function of integration.

Interface   A circuit or device, e.g. a modem, that enables a computer to transfer data to and from its surroundings or between computers.

Inverting amplifier   An amplifier whose output is 180 degrees out of phase with its input.

Ion   An atom or group of atoms that has gained or lost one or more electrons, and which therefore carries a positive charge.

Isothermal block   A connecting block used with thermocouples.

Jack   A connecting device which is arranged for the insertion of a plug to which the wires of a circuit may be connected.

Joule   The SI unit of energy.

Junction   A region of contact between two dissimilar metals (as in a thermocouple) or two dissimilar conductors (as in a diode) which has useful electrical properties.

Kilobit   One thousand bits, i.e. 0s and 1s of data.

Kilobyte   One thousand bytes of data.

Kilohertz (kHz)   A frequency equal to 1000 Hz.

Large-scale integration (LSI)   The process of making integrated circuits with between 100 and 5000 logic gates on a single silicon chip.

Laser   A device that produces an intense and narrow beam of light of almost one particular wavelength. The light from lasers is used in optical communications systems, compact disc players and video disc players.

LCD (liquid crystal display)   A display that operates by controlling the reflected light from special liquid crystals, rather than by emitting light as in the light-emitting diode.

LDR (light-dependent resistor)   A semiconductor device that has a resistance decreasing sharply with increasing light intensity. The LDR is used in light control and measurement systems, e.g. automatic street lights and cameras.

LED (light-emitting diode)   A small semiconductor diode that emits light when current passes between its anode and cathode terminals. Red, green, yellow and blue LEDs are used in all types of display systems, e.g. hi-fi amplifiers.

Limit switch   A switch that is arranged to be actuated by a workpiece.

Load   The general name for a device e.g. an electric motor, that absorbs electrical energy.

Load cell   A device for measuring weight. Weight resting on the device causes compression strain. Weight suspended from the device causes tensile strain. Strain is reported as a change in resistance by a coupled strain gauge.

Logic circuit   An electronic circuit that carries out simple logic functions.

Logic diagram   A circuit diagram showing how logic gates and other digital devices are connected together to produce a working circuit or system.

Logic gate   A digital device e.g. an AND gate, that produces an output of logic 1 or 0 depending on the combination of 1s and 0s at its inputs.

Loudspeaker   A device used to convert electrical energy into sound energy.

Low pass filter   A filter that passes all frequencies below a specific frequency.

LVDT (linear variable differential transformer)   A device used for position detection.

Machine code   Instructions in the form of patterns of binary digits which enable a computer to carry out calculations.

Magnetic bubble memory (MBM)   A device that stores data as a string of magnetic 'bubbles' in a thin film of magnetic material. The MBM can store a very large amount of data in small volume and is ideal for portable computer products such as word processors.

Magnetic reed switch   A magnetically operated switch. Made of two or three magnetic leaves in a glass tube. Proximity of a magnet causes the switch to close.

Magnetic storage   Magnetic tapes, floppy discs and magnetic bubble memories that store data as local changes in the strength of a magnetic field, and which can be recovered electrically.

Majority carrier   The most abundant of the two charge carriers present in a conductor. The majority charge carriers in n-type material are electrons.

Man–machine interface   Any hardware, e.g. a keyboard or mouse, that allows a person to exchange information with a computer or machine.

Mark-to-space ratio   The ratio of the time that the waveform of a rectangular waveform is *high* to the time it is *low*.

Mass flowmeter   A fluid-flow measuring device that measures the mass of the fluid instead of its velocity. Used when great accuracy is required.

Matrix   A logical network in the form of a rectangular array of intersections.

Medium waves   Radio waves having wavelengths in the range about 200 to 700 m, i.e. frequencies in the range 1.5 to 4.5 MHz.

Megabit   A quantity of data equivalent to one million (10^6) bits.

Megabyte   A quantity of binary data equal to one million (10^6) bytes. Floppy discs store approximately this amount of data.

Memory   That part of a computer system used for storing data until it is needed. A microprocessor in a computer can locate and read each item of data by using an address.

Memory-mapped interface   An interface system in which the input/output ports are addressed as memory locations.

Microcomputer   A usually portable computer which can be programmed to perform a large number of functions quickly and relatively cheaply. Its main uses are in the home, school, laboratory and office.

Microelectronics   The production and use of complex circuits on silicon chips.

Microfarad   A unit of electrical capacitance equal to one millionth of a farad.

Micron (micrometre)   A distance equal to one millionth of a metre. The micron is used for measuring the size and separation of components on silicon chips.

Microprocessor   A complex integrated circuit manufactured on a single silicon chip. It is the 'heart' of a computer and can be programmed to perform a wide range of functions. A microprocessor is used in washing machines, cars, cookers, games and many other products.

Microswitch   A small mechanically operated switch.

Microwaves   Radio waves having wavelengths less than about 300 mm and used for straight line communications by British Telecom and others.

Minority carrier   The least abundant of the two charge carriers present in a semiconductor. The minority charge carriers in n-type material are holes.

Modem (modulator/demodulator)   A device for converting computer data in digital form into analogue signals for transmission down a telephone.

Modulator   A circuit that puts a message on some form of carrier wave.

Monostable   A circuit that produces a time delay when it is triggered, and then reverts back to its original, normally stable, state.

Mouse   A small hand-operated device connected to a computer by a trailing wire, or by optical means, that makes a cursor move around the screen of a VDU to select operations and make decisions.

MSB (most significant bit)   The left-hand binary digit in a digital word.

Multimeter   An instrument for measuring current, potential difference and resistance, and used for testing and fault-finding in the design and use of electronic circuits.

Multiplexing   A method of making a single communications channel carry several messages.

Multivibrator   Any one of three basic types of two-stage transistor circuit in which the output of each stage is fed back to the input of the other stage using coupling capacitors and resistors, and causing the transistors to switch on and off rapidly. The multivibrator family includes the monostable, astable and bistable.

Negative feedback   The feeding back to the input of a system a part of its output signal. Negative feedback reduces the overall gain of an amplifier but increases its bandwidth and stability.

Neutron   A particle in the nucleus of an atom which has no electrical charge and a mass roughly equal to that of the proton.

Noise   An undesirable electrical disturbance or interference.

Noise (white)   Noise which is made up of a frequency spectrum (like white light).

Node   A point of zero voltage or zero current on a conductor or the point in a radio wave where the amplitude is zero.

Nucleus   The central and relatively small part of an atom that is made up of protons and neutrons.

Open loop control   A control system in which no self-correcting action occurs as it does in a closed loop system.

Open loop gain   The gain of an amplifier without feedback.

Operational amplifier (op. amp)   A very high gain amplifier that produces an output voltage proportional to the difference between its two input voltages. Op.amps are widely used in instrumentation and control systems.

Optical fibre   A thin glass or plastic thread through which light travels without escaping from its surface.

Optoelectronics   A branch of electronics dealing with the interaction between light and electricity. Light-emitting diodes and liquid crystal displays are examples of optoelectronic devices.

Oscillator   A circuit or device, e.g. an audio frequency oscillator, that provides a sinusoidal or square wave voltage output at a chosen frequency. An astable multivibrator is one type of oscillator.

Package   The plastic or ceramic material used to cover and protect an integrated circuit.

Parabolic reflector   A hollow concave reflector.

Passband   The range of frequencies passed by a filter.

Passive filter   A filter made of passive components: resistors, capacitors and inductors.

PCB (printed circuit board   A thin board made of electrically insulating material (usually glass fibre) on which a network of copper tracks is formed to provide connections between components soldered to the tracks.

Period   The time taken for a wave to make one complete oscillation. The period of the 50 Hz mains frequency is 0.02 s.

Photodiode   A light-sensitive diode that is operated in reverse bias. When light strikes the junction the diode goes into reverse breakdown and conducts. It is able to respond rapidly to changes of light.

Photoelectron   An electron released from the surface of a metal by the action of light.

Photomask   A transparent glass plate used in the manufacture of integrated circuits on a silicon chip.

Photon   The smallest 'packet', or quantum, of light energy.

Photoresist   A light-sensitive material that is spread over the surface of a silicon wafer from which silicon chips are made.

Photoresistor   A transistor that responds to light and produces an amplified output signal. Like photodiodes, photoresistors respond rapidly to light changes and are used as sensors in optical communications systems.

Photovoltaic   The property of responding to light with an electrical current. Photovoltaic cells are used in generating electricity from solar energy.

Picofarad (pF)   An electrical capacitance equal to one millionth of a farad (10^{-12} F)

Piezoelectricity   The electricity that crystals, such as quartz, produce when they are squeezed. Conversely, if a potential difference is applied across a piezoelectric crystal, it alters shape slightly. The piezoelectric effect is used in digital watches, hi-fi pick-ups and gas lighters.

Plasma   A completely ionised gas at extremely high temperatures.

Port   A place on a microcomputer to which peripherals can be connected to provide two-way communication between the computer and the outside world.

Positive displacement flowmeter   A device that measures fluid flow by passing the fluid in measured increments. Usually accomplished by alternately filling and emptying a chamber.

Positive feedback   The feeding back to the input of a system a part of its output signal. Positive feedback increases the overall gain of an amplifier and is used in an astable multivibrator.

Potential divider   Two or more resistors connected in series through which current flows to produce potential differences dependent on the resistor values.

Potentiometer   An electrical component, having three terminals, that provides an adjustable potential difference.

Power   The rate of doing work. Measured in watts.

Preferred value   Manufacturers' standardised component values used in resistor and capacitor values.

Programme   A set of instructions used for the collation of data or for the solution of a problem.

Proton   A particle that makes up the nucleus of an atom and has a positive charge equal in value to the negative charge of the electrons.

Pulse   A short-lived variation of voltage or current in a circuit

Q-factor   The sharpness (or 'quality') of an electronic filter circuit, e.g. a tuned circuit, that enables it to accept or reject a particular frequency.

Quantum   The smallest packet of radiant energy, e.g. a photon, that can be transmitted from place to place and described by Planck's quantum theory.

Quartz   A crystalline form of silicon dioxide which has piezoelectric properties and can, therefore, be used as a pressure transducer and to provide a stable frequency in, for example, crystal clocks.

Qwerty keyboard   A keyboard (e.g. a computer keyboard) that has its keys arranged in the same way as those of a standard typewriter, i.e. the first six letters of the top row spell 'QWERTY'.

Radiation   Energy travelling in the form of electromagnetic waves.

Radio   The use of electromagnetic waves to transmit or receive electrical signals without connecting wires.

RAM (random access memory)   An integrated circuit that is used for the temporary storage of computer programs.

Rectifier   A semiconductor diode that makes use of the one-way conducting properties of a pn junction to convert a.c. to d.c.

Relay   A magnetically operated switch that enables a small current to control a much larger current in a separate circuit.

Resistance   The opposition offered by a component to the passage of electricity.

Resonance   The build-up of large amplitude oscillations in a tuned circuit.

Response time   The time required for a system to return to normal following a disturbance.

Reverse bias   A voltage applied across a pn junction (e.g. a diode) which prevents the flow of electrons across the junction.

Rheostat   An adjustable resistor.

rms (root mean square) value   The value of an alternating current which has the same heating effect as a steady d.c. current. 240 V is the rms value of the mains voltage.

Robot   A computer-controlled device that can be programmed to perform repetitive tasks such as paint-spraying, welding and machining of parts.

ROM (read-only memory)   An integrated circuit that is used for holding data permanently, e.g. for storing the language and graphics symbols used by a computer.

Schmitt trigger   A snap-action electronic switch which is widely used to 'sharpen up' slowly changing waveforms.

SCR (silicon-controlled rectifier)   A four-layer semiconductor device used in switching circuits. Also known as a thyristor.

Semiconductor   A solid material that is a better electrical conductor than an insulator but not such a good conductor as a metal. Diodes, transistors and integrated circuits are based on n-type and p-type semiconductors.

Sensor   Any device which produces an electrical signal indicating a change in its surroundings.

Sequential logic   A digital circuit that can store information. Sequential logic circuits are based on flip-flops and are the basis of counters and computer memories.

Servosystem   An electromechanical system which uses sensors to control and monitor precisely the movement of something.

Short waves   Radio waves that have wavelengths between about 2.5 MHz and 15 MHz and which are mainly used for amateur and long-range communications.

Silicon   An abundant non-metallic element used for making diodes and transistors. Silicon is doped with small amounts of impurities such as boron and phosphorus to make n-type and p-type semiconductors.

Silicon chip   A small piece of silicon on which a

complex miniaturised circuit (called an integrated circuit) is formed by photographic and chemical processes.

Small-scale integration (SSI) The process of making integrated circuits.

Software Instructions or programs stored in a computer system.

Solder An alloy of tin and lead that has a low melting point and is used for making electrical connections between components on a circuit board.

Solenoid A coil of copper wire in which an iron rod moves by the magnetic field produced when a current flows through the coil.

Stepping motor An electric motor with a shaft that rotates one step at a time. Stepping (or stepper) motors are used for the precise positioning of robot arms.

Strain Change in dimension of a material when force is applied.

Strain gauge A device used to measure strain. The change in electrical resistance of the strain gauge is a measure of the strain.

Summing amplifier An amplifier whose output is proportional to the sum of two or more input signals, or an amplifier used to add (sum) its input voltages.

System All the parts which make up a working whole.

Tachogenerator A device used to measure motor speed. The output is a voltage or a frequency that is proportional to motor speed.

Telex An audio frequency teleprinter system provided by the Post Office for use over telephone lines.

Tensile strain Strain caused by force pulling on a member.

Thermistor A semiconductor temperature sensor.

Thermocouple A temperature-sensing device whose output is a current or voltage which is proportional to the difference between the temperatures at two junctions of dissimilar metals.

Thermopile A system of several thermocouples in a series-aiding configuration. This configuration increases the sensitivity of the thermocouple.

Thyristor A half-wave semiconductor switching device used for motor speed control and lamp dimming. Also known as a silicon-controlled rectifier (SCR).

Time constant The time taken for the voltage across a capacitor, to rise to 63% of its final voltage when it charges through a resistor connected in series with it.

Torque Twisting or rotary force, such as that delivered by a motor shaft.

Transducer A device which converts mechanical or physical quantities into electrical quantities, or a device which converts electrical quantities into physical quantities.

Transformer An electromagnetic device for converting alternating current from one voltage to another.

Transistor A semiconductor device which has three terminals and is used for switching and amplification.

Transmitter A device or equipment which converts audio or video signals into modulated radio frequency signals which are then sent (transmitted) by electromagnetic waves.

TRIAC A full-wave semiconductor switching device used for motor speed control and lamp dimming.

Truth table A list of 0s and 1s that shows how a digital logic circuit responds to all possible combinations of binary input signals.

TTL (transistor-transistor logic) The most common type of IC logic in use today.

Tuned circuit A circuit which contains an inductor and a capacitor and can be tuned to receive particular radio signals.

Tweeter A loudspeaker used to reproduce the higher audio frequencies (above 5 kHz).

UHF (ultra-high frequency) Radio waves that have frequencies in the range 500 MHz to 30,000 MHz and are used for TV broadcasts.

UJT (uni-junction transistor) A type of transistor used as a relaxation oscillator in SCR control circuits.

Unipolar transistor A transistor that depends for its operation on either n-type or p-type semiconductor materials as in a field-effect transistor.

Ultrasonic waves Sound waves inaudible to the human ear that have frequencies above about 20 kHz.

Ultraviolet Radiation having wavelengths between the visible violet and the X-ray region of the electromagnetic spectrum.

VDU (video display unit) An input/output device comprising a screen and sometimes a keyboard that enables a person to communicate with a computer.

Velocity flowmeter A device that measures fluid flow directly. The most common is the turbine flowmeter.

VHF (very high frequency) Radio waves that have frequencies in the range 30 MHz to 300 MHz and are used for high-quality radio broadcasts (FM) and TV transmission.

Viewdata An information service that enables telephone subscribers to access a wide range of information held in a database and which is displayed on a TV set coupled to the telephone line by a modem.

VMOS (vertical metal-oxide semiconductor) A type of field-effect transistor. It is a small high-power fast-acting transistor used in audio amplifiers and power switching circuits.

Voltage difference amplifier An amplifier whose output is proportional to the difference between two input voltages.

Wafer A thin disc cut from a single crystal of silicon on which hundreds of integrated circuits are made before being cut up into individual ICs for packaging.

Waveform The shape of an electrical signal, e.g. a sinusoidal waveform.

Wavelength The distance between one point on a wave and the next corresponding point.

Wheatstone bridge   A network of resistors used to measure very small changes in resistance.

Woofer   A loudspeaker used to reproduce the lower audio frequencies.

Word   A pattern of bits (i.e. 1s and 0s) that is handled as a single unit of information in digital systems; e.g. a byte is an 8-bit word.

Wordprocessor   A computerised typewriter that allows written material to be generated, stored, edited, printed and transmitted.

X-axis deflection   Horizontal deflection on the screen of a CRO, often used as the time base.

X-rays   Penetrating electromagnetic radiation used in industry and in medicine for seeing below the surface of solid materials.

Y-axis deflection   Vertical deflection on the screen of a CRO.

Zener diode   A special semiconductor diode that is designed to conduct current in the reverse-bias direction at a particular reverse-bias voltage. Zener diodes are widely used to provide stabilised voltages in electronic circuits.

INDEX